GUIDED READING ACTIVITIES
WITH ANSWER KEY

People, Places, AND Change

HOLT, RINEHART AND WINSTON
A Harcourt Classroom Education Company

Austin • New York • Orlando • Atlanta • San Francisco • Boston • Dallas • Toronto • London

Cover: Hilary Wilkes/International Stock Photography

Front and Back Cover background, and Title Page: Artwork by Nio Graphics. Rendering based on photo by Stone/Cosmo and Condina.

ISBN 0-03-054868-3

1 2 3 4 5 6 7 8 9   082   04 03 02 01 00

TABLE OF CONTENTS

TABLE OF CONTENTS

A Geographer's World

SECTION 1

Reading the Section • As you read the section, complete the following outline by supplying the main idea and the missing subtopics and supporting details.

Developing a Geographic Eye

Main Idea: _____

_____

Topic I: The study of geography is important.

   Detail A: People familiar with geography see meaning in the arrangement of things on Earth.

   Detail B: _____

   Detail C: People familiar with geography can understand the world around them.

Topic II: _____

   Detail A: Geographers study Earth's processes and their impact on people.

   Detail B: _____

Topic III: Geographers study issues at different levels.

   Detail A: Studying geography at the local level is a good way to get to know a community.

   Detail B: _____

   Detail C: _____

Post-Reading Quick Check • After you have finished reading the section, in the space provided, explain in your own words why the study of geography is important to you.

_____

_____

_____

_____

_____

Name _____ Class _____ Date _____

A Geographer's World

Reading the Section • As you read the section, complete the graphic organizer by identifying and describing the five themes of geography.

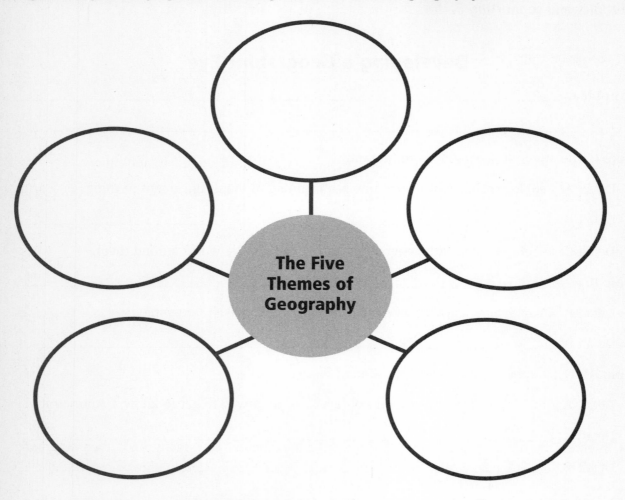

The Five Themes of Geography

Post-Reading Quick Check • After you have finished reading the section, in the space provided, identify three ways in which people interact with the environment.

1. _____

2. _____

3. _____

People, Places, and Change

CHAPTER 1

A Geographer's World

Reading the Section • As you read the section, examine the descriptions of the branches of geography below. For each description, place a check mark in the box next to the branch of geography being described.

1. Focuses on the art and science of mapmaking
❑ Human geography   ❑ Physical geography   ❑ Cartography   ❑ Meteorology   ❑ Climatology

2. Branch of geography that tracks Earth's larger atmospheric systems
❑ Human geography   ❑ Physical geography   ❑ Cartography   ❑ Meteorology   ❑ Climatology

3. People who work in this field forecast and report temperature, rainfall, and other atmospheric conditions
❑ Human geography   ❑ Physical geography   ❑ Cartography   ❑ Meteorology   ❑ Climatology

4. Focuses on the study of people—past or present, their location and distribution over Earth, their activities, and their differences
❑ Human geography   ❑ Physical geography   ❑ Cartography   ❑ Meteorology   ❑ Climatology

5. People in this field study landforms such as mountains, plains, and deserts
❑ Human geography   ❑ Physical geography   ❑ Cartography   ❑ Meteorology   ❑ Climatology

6. Focuses on Earth's natural landscapes and physical systems
❑ Human geography   ❑ Physical geography   ❑ Cartography   ❑ Meteorology   ❑ Climatology

7. Most people in this branch of geography do their work on computers
❑ Human geography   ❑ Physical geography   ❑ Cartography   ❑ Meteorology   ❑ Climatology

8. Subbranches of this field include political geography, economic geography, and urban geography
❑ Human geography   ❑ Physical geography   ❑ Cartography   ❑ Meteorology   ❑ Climatology

Post-Reading Quick Check • After you have finished reading the section, in the space provided, identify five occupations in which people use geography.

1. Occupation: _____

2. Occupation: _____

3. Occupation: _____

4. Occupation: _____

5. Occupation: _____

CHAPTER 2

Planet Earth

Reading the Section • As you read the section, examine each of the pairs of statements below. Circle the letter of the statement in each pair that is true.

1. **a.** The solar system consists of Earth and the objects that move around it.
   b. The solar system consists of the Sun and the objects that move around it.

2. **a.** The planet nearest the Sun is Earth.
   b. The planet nearest the Sun is Mercury.

3. **a.** A satellite is a body that orbits a larger body.
   b. An equinox is a body that orbits a larger body.

4. **a.** The diameter of the Sun is about 100 times the diameter of Earth.
   b. The diameter of Earth is about 100 times the diameter of the Sun.

5. **a.** One complete spin of Earth on its axis is called a revolution.
   b. One complete spin of Earth on its axis is called a rotation.

6. **a.** It takes 10 years for Earth to complete one orbit around the Sun.
   b. It takes one year for Earth to complete one orbit around the Sun.

7. **a.** Earth's axis is tilted, or slanted.
   b. Earth's axis is straight up and down.

8. **a.** The Arctic Circle is the line of latitude located 66.5° north of the equator.
   b. The Arctic Circle is the line of latitude located 66.5° south of the equator.

9. **a.** The Tropic of Capricorn is the line of latitude 23.5° south of the equator.
   b. The Tropic of Capricorn is the line of latitude 23.5° north of the equator.

10. **a.** During a solstice, every place on Earth has 12 hours of day and 12 hours of night.
    b. During an equinox, every place on Earth has 12 hours of day and 12 hours of night.

Post-Reading Quick Check • After you have finished reading the section, in the space provided, identify and describe the four parts of the Earth system.

1. Part: _____ Description: _____

2. Part: _____ Description: _____

3. Part: _____ Description: _____

4. Part: _____ Description: _____

Planet Earth

Reading the Section • As you read the section, answer each of the following questions in the space provided.

1. What do we call water in the form of an invisible gas? _____

2. What is the water cycle? _____

3. The oceans contain what percentage of Earth's water? _____

4. Why do some states create reservoirs? _____

5. What is a tributary? _____

6. What do we call water that fills the spaces between soil and grains of rock? _____

7. What are the names of Earth's four oceans? _____

8. Where does marine life tend to concentrate? _____

9. What is the world's deadliest natural hazard? _____

10. What is the purpose of building dams? _____

Post-Reading Quick Check • After you have finished reading the section, in the space provided, describe the water cycle.

_____

_____

_____

_____

_____

_____

_____

_____

_____

_____

Planet Earth

CHAPTER 2

Reading the Section • As you read the section, match each term in the left column with its description in the right column. Write the letter of the correct term in the space provided.

_____ **1.** core

_____ **2.** lava

_____ **3.** continent

_____ **4.** earthquake

_____ **5.** mid-ocean ridge

_____ **6.** erosion

_____ **7.** Pangaea

_____ **8.** alluvial fan

_____ **9.** delta

_____ **10.** glacier

a. Landform similar to an underwater mountain range

b. Large, slow-moving river of ice

c. Earth's original supercontinent

d. Inner, solid part of Earth

e. Sediment deposited into a fan-shaped pile

f. One of Earth's large landmasses

g. Landform found where the river joins the ocean

h. Magma that has reached the surface of Earth

i. Movement of rocky materials to another location

j. Sudden, violent movement along a fracture within Earth's crust

Post-Reading Quick Check • After you have finished reading the section, in the space provided, identify three ways that Earth's plates move, and describe what happens when they move.

1. Movement: _____ What happens: _____

_____

2. Movement: _____ What happens: _____

_____

3. Movement: _____ What happens: _____

_____

Name _____ Class _____ Date _____

Wind, Climate, and Natural Environments

CHAPTER 3

Reading the Section • As you read the section, circle the boldface word or phrase that *best* completes each statement below.

1. All planets in our solar system receive energy from the **Sun / Moon**.

2. When the direct rays of the Sun strike Earth at the **Tropic of Capricorn / Tropic of Cancer**, it is summer in the Northern Hemisphere.

3. From year to year, temperatures in any one place usually **vary / stay the same**.

4. The **greenhouse effect / hothouse effect** refers to the process by which the atmosphere traps heat.

5. The term air pressure refers to the **weight / moisture** of the air.

6. Air pressure is measured by using a **thermometer / barometer**.

7. When air warms, it gets **lighter / heavier** and **falls / rises**.

8. Wind travels from areas of **low / high** pressure to areas of **low / high** pressure.

9. When a large amount of warm air meets a large amount of cold air, an area of unstable weather, called a **front / back**, forms.

10. The Gulf Stream is an **ocean / air** current.

Post-Reading Quick Check • After you have finished reading the section, in the space provided, identify and describe four types of winds.

1. Type: _____ Description: _____

   _____

2. Type: _____ Description: _____

   _____

3. Type: _____ Description: _____

   _____

4. Type: _____ Description: _____

   _____

Wind, Climate, and Natural Environments

CHAPTER 3

SECTION 2

Reading the Section • As you read the section, use the space provided to identify the type of climate described by each statement. Choose your answers from the list below. Some answers will not be used.

humid tropical	Mediterranean	subarctic
tropical savanna	humid subtropical	tundra
desert	marine west coast	ice cap
steppe	humid continental	highland

_____ **1.** This climate results in many barren, rocky, or sandy areas.

_____ **2.** Temperate evergreen forests can be found in this climate region.

_____ **3.** The polar regions of Earth have this type of climate.

_____ **4.** Vegetation in this climate region is grass with scattered trees and shrubs.

_____ **5.** This climate region has long, cold winters, short summers, and little rain.

_____ **6.** The tropical rain forest exists in this climate region.

_____ **7.** This climate can be found in the southeastern United States.

_____ **8.** Permafrost can be found in this climate region.

_____ **9.** Most of the variable weather in this climate region is the result of cold and warm air coming together.

_____ **10.** This climate is found between desert and wet climate regions.

Post-Reading Quick Check • After you have finished reading the section, in the space provided, explain the difference between weather and climate.

_____

_____

_____

_____

_____

Wind, Climate, and Natural Environments

CHAPTER 3

Reading the Section • As you read the section, complete each sentence below by writing the appropriate word or phrase in the space provided.

1. Environmental changes that are too great may cause some types of plants or animals

   to become _____, or die out completely.

2. _____ refers to the study of the connections among different forms of life.

3. The process by which plants convert sunlight into chemical energy is known as

   _____.

4. _____ take in minerals, water, and gases from the soil.

5. Any type of plant or animal tends to be most common in the setting where it is

   best able to live, grow, and _____.

6. Soil _____ are substances in the soil that promote growth.

7. Plant _____ are groups of plants that live together in the same area.

8. All of the plants and animals in an area together with the nonliving parts of their

   environment form what is known as a(n) _____.

9. Plant _____ refers to the gradual process by which one group of plants replaces another group of plants.

10. Soils that are rich in _____, or decayed plant and animal matter, are described as fertile.

Post-Reading Quick Check • After you have finished reading the section, in the space provided, describe the food chain.

_____

_____

_____

_____

_____

Earth's Resources

SECTION 1

Reading the Section • As you read the section, complete the following outline by supplying the main idea and the missing subtopics and supporting details.

Soil and Forests

Main Idea: _____

_____

Topic I: Protecting Earth's soil can be challenging.

Detail A: Farmers fertilize the soil or use crop rotation to enhance and protect the soil.

Detail B: _____

Detail C: Farmers plant rows of trees or build terraces into slopes to prevent erosion.

Detail D: _____

Topic II: _____

Detail A: Two ways to protect forests are legislation and reforestation.

Detail B: _____

Post-Reading Quick Check • After you have finished reading the section, in the space provided, identify eight products that are provided by forests.

1. Product: _____

2. Product: _____

3. Product: _____

4. Product: _____

5. Product: _____

6. Product: _____

7. Product: _____

8. Product: _____

People, Places, and Change

CHAPTER 4 Earth's Resources

Reading the Section • As you read the section, examine the riddles below. Solve each riddle by writing the correct word or words in the space provided.

_____ **1.** "Because I am a place that receives only a small amount of rain, I am better suited to grazing animals than to farming. What kind of place am I?"

_____ **2.** "People who live in dry regions often build me, along with canals and reservoirs, to bring in fresh water. What am I?"

_____ **3.** "I live deep under the ground and hold a great deal of water. One of my relatives, Ogallala, is so big that he stretches from Texas to South Dakota. What am I?"

_____ **4.** "Even though I am an expensive process, people seem to like the way I remove salt from seawater. What process am I?"

_____ **5.** "Because I do not need to be watered as often as those silly, thirsty grasses, people who live in dry climates like to use me for landscaping. What am I?"

_____ **6.** "I am formed when air pollution combines with moisture in the air and then falls to the ground. I can be quite damaging to trees and things that live in the sea. What am I?"

_____ **7.** "I am the layer in the upper atmosphere that protects living things by absorbing most of the Sun's harmful ultraviolet light. Damage me and you damage yourselves! What am I?"

_____ **8.** "I am the term scientists use for the slow increase in Earth's average temperature. What term am I?"

Post-Reading Quick Check • After you have finished reading the section, in the space provided, explain how people pollute the water supply.

_____

_____

_____

_____

_____

Name _____ Class _____ Date _____

Earth's Resources

Reading the Section • As you read the section, complete the chart below by writing each of the following minerals under the correct heading.

sapphire	aluminum	talc	copper
silver	ruby	gold	emerald
sulfur	quartz	salt	platinum
iron	steel	mercury	diamond

Metallic Minerals	**Nonmetallic Minerals**

Post-Reading Quick Check • After you have finished reading the section, in the space provided, define the term *mineral* and list the four properties of minerals.

1. Definition: _____

_____

2. Property: _____

3. Property: _____

4. Property: _____

5. Property: _____

People, Places, and Change

Earth's Resources

Reading the Section • Each of the following sentences contains an underlined word or phrase that makes the sentence incorrect. As you read the section, use the space provided to write the word or phrase that makes the sentence correct.

_____ 1. Fossil fuels such as coal, petroleum, and natural gas are <u>renewable resources</u>.

_____ 2. Until the 1900s people for the most part used wood and <u>hydroelectric power</u> as sources of energy.

_____ 3. When it is first pumped out of the ground, petroleum is called <u>refined</u> oil.

_____ 4. More than half of the world's oil that we know about can be found in <u>North America</u>.

_____ 5. <u>Coal</u> is the cleanest-burning fossil fuel.

_____ 6. The renewable energy resource that is the most widely used is <u>geothermal energy</u>.

_____ 7. Dams produce <u>50</u> percent of the electricity used in the United States.

_____ 8. Earth's internal heat, called <u>solar energy</u>, escapes through hot springs and steam vents on Earth's surface.

_____ 9. <u>Nuclear power</u> comes from the Sun's heat and light.

_____ 10. In 1986 there was a nuclear accident in the city of <u>Kiev</u> in the Ukraine that killed a number of people.

Post-Reading Quick Check • After you have finished reading the section, in the space provided, explain the difference between renewable and nonrenewable resources.

_____

_____

_____

_____

_____

The World's People

SECTION 1

Reading the Section • As you read the section, examine each of the pairs of statements below. Circle the letter of the statement in each pair that is true.

1. **a.** Another term for culture is ethnic group.
   b. Culture is a learned system of shared beliefs and ways of doing things that guide daily behavior.

2. **a.** Race is based on inherited physical or biological traits.
   b. Race is based on such things as language and religion.

3. **a.** Cultures change over time.
   b. Cultures always stay the same.

4. **a.** Diffusion refers to the tendency of cultures to resist change.
   b. Diffusion refers to the spread of one culture's ways to another culture.

5. **a.** Two important factors influencing the way people meet basic needs are their history and environment.
   b. Two important factors influencing how people meet basic needs are race and religion.

6. **a.** American cities are oriented to the four compass points.
   b. American cities often follow a rectangular grid plan.

7. **a.** A domesticated species can never be tamed by humans.
   b. A domesticated species has changed so much that it depends on humans to survive.

8. **a.** A civilization is a highly complex culture.
   b. The term *civilization* refers to a very simple culture.

Post-Reading Quick Check • After you have finished reading the section, in the space provided, explain how symbols contribute to cultural differences.

_____

_____

_____

_____

_____

The World's People

Reading the Section • As you read the section, circle the boldface word or phrase that *best* completes each statement below.

1. The study of human populations is called **demography / climatology**.

2. A country's **population density / birthrate** is measured by dividing the number of people in the country by the area of that country.

3. **Ruralization / Urbanization** refers to the movement of people from farms to cities.

4. The **birthrate / death rate** minus the **birthrate / death rate** equals the rate of natural increase in a country.

5. People who work in **tertiary / quaternary** industries have specialized skills or knowledge.

6. The **literacy rate / gross national product** is a common way to measure a country's economy.

7. The United States, Canada, Japan, and most countries in Europe are known as **developed countries / developing countries**.

8. Under a **communist / democratic** system of government, voters elect leaders and rule by majority.

Post-Reading Quick Check • After you have finished reading the section, in the space provided, discuss the characteristics shared by developing countries.

_____

_____

_____

_____

_____

_____

_____

_____

The World's People

CHAPTER 5

SECTION 3

Reading the Section • As you read the section, complete the chart below about Earth's future population growth by providing arguments that support each point of view.

Earth <u>Can</u> Support a Much Larger Human Population	Earth <u>Cannot</u> Support a Much Larger Human Population

Post-Reading Quick Check • After you have finished reading the section, in the space provided, list two problems associated with a high population growth rate and two problems associated with a low population growth rate.

1. Problem of high growth rate: _____

_____

2. Problem of high growth rate: _____

_____

3. Problem of low growth rate: _____

_____

4. Problem of low growth rate: _____

_____

CHAPTER 6 — The United States

Reading the Section • As you read the section, examine the descriptions of U.S. regions below. For each description, place a check mark in the box next to the region being described.

1. The Mississippi River and its tributaries drain this region of the United States.
   ❑ The East   ❑ The Interior   ❑ The West

2. This region contains the Rockies and a region of basins and plateaus.
   ❑ The East   ❑ The Interior   ❑ The West

3. This region rises from the Coastal Plains to the Appalachian Mountains.
   ❑ The East   ❑ The Interior   ❑ The West

4. Near to the Pacific coast are the Sierra Nevada, the Coast Ranges, and valleys.
   ❑ The East   ❑ The Interior   ❑ The West

5. The flattest part of this region is the Great Plains region.
   ❑ The East   ❑ The Interior   ❑ The West

6. Dry climates are most common in this region.
   ❑ The East   ❑ The Interior   ❑ The West

7. The northern part of the West Coast has a marine west coast climate.
   ❑ The East   ❑ The Interior   ❑ The West

8. Southern Florida, which has a tropical savanna climate, is warm all year round.
   ❑ The East   ❑ The Interior   ❑ The West

9. The steppe climate of the Great Plains supports wide grasslands.
   ❑ The East   ❑ The Interior   ❑ The West

10. People in this region sometimes experience hail and tornadoes.
    ❑ The East   ❑ The Interior   ❑ The West

Post-Reading Quick Check • After you have finished reading the section, in the space provided, discuss the natural resources of the United States.

_____

_____

_____

_____

The United States

Reading the Section • As you read the section, examine the
riddles below. Solve each riddle by writing the correct word or
words in the space provided.

_____ **1.** "About 20,000 years ago, some of my people left me and moved to
North America. How brave of them! What country am I?"

_____ **2.** "After around A.D. 700, we developed a complex irrigation system in
the southwestern United States. Who are we?"

_____ **3.** "I am a large farm that grows mainly one crop. I was especially
common in the southern British colonies. What am I?"

_____ **4.** "In 1776 my 13 colonies in North America decided to break away
from me. Talk about ungrateful! What country am I?"

_____ **5.** "The United States annexed me to the country in 1898. Aloha! What
state am I?"

_____ **6.** "I have been the basis of the U.S. government since the late 1780s. I
am so good that other countries use me as a model. What am I?"

_____ **7.** "When people mention me, they are referring to the ability to speak
two languages. What term am I?

_____ **8.** "I am Independence Day in the United States. Americans celebrate
me with fireworks and picnics. What date am I?"

Post-Reading Quick Check • After you have finished reading
the section, in the space provided, identify four religious holidays
celebrated in the United States.

1. Holiday: _____

2. Holiday: _____

3. Holiday: _____

4. Holiday: _____

CHAPTER 6 — The United States

Reading the Section • All of the following statements refer to the economy of the U.S. regions. As you read the section, use the space provided to indicate whether each statement below refers to the Northeast, the South, the Midwest, the Interior West, or the Pacific.

_____ **1.** In this region today, ranching is often combined with wheat farming.

_____ **2.** Historically, this region was rural and agricultural.

_____ **3.** Fish and forests are important resources in this region's northwest states.

_____ **4.** Most farms here are small because of a short growing season and rocky terrain.

_____ **5.** Good soils and flat land have helped make this region one of the world's great farming regions.

_____ **6.** This region was the country's first industrial area.

_____ **7.** Hawaii, found in this region, has natural beauty and a mild climate.

_____ **8.** This region is home to the Corn Belt and the Dairy Belt.

_____ **9.** Today, many textile factories are located in this region's Piedmont areas.

_____ **10.** This region has rich deposits of coal, oil, gold, silver, and copper.

Post-Reading Quick Check • After you have finished reading the section, in the space provided, identify six challenges facing the United States today.

1. Challenge: _____

2. Challenge: _____

3. Challenge: _____

4. Challenge: _____

5. Challenge: _____

6. Challenge: _____

Canada

Reading the Section • As you read the section, answer each of the following questions in the space provided.

1. With what country does Canada share major physical regions? _____

2. Which mountains extend into southeastern Canada? _____

3. What river links the Great Lakes to the Atlantic Ocean? _____

4. What carved out many of Canada's lakes? _____

5. Where can the mildest part of Canada be found? _____

6. What two climate types does the far north of Canada have? _____

7. About what percentage of Canada contains permafrost? _____

8. What Canadian resource stretches from Labrador to the Pacific coast? _____

9. What do Canadians make from tree pulp? _____

10. Which of Canada's resources is the most valuable? _____

Post-Reading Quick Check • After you have finished reading the section, in the space provided, describe the mineral resources of Canada.

_____

_____

_____

_____

_____

_____

_____

_____

_____

_____

People, Places, and Change

Name _____ Class _____ Date _____

CHAPTER 7 Canada

Reading the Section • As you read the section, write the letter of the *best* choice in the space provided.

_____ **1.** Canada's two official languages are
   a. Spanish and French.
   b. Portuguese and Spanish.
   c. English and French.
   d. English and Spanish.

_____ **2.** The first Europeans to settle in Canada probably were the
   a. Vikings.
   b. Swiss.
   c. British.
   d. Russians.

_____ **3.** What percentage of present-day Canadians are of French ancestry?
   a. nearly 25 percent
   b. slightly more than 50 percent
   c. around 75 percent
   d. slightly less than 100 percent

_____ **4.** Which of the following groups first came to Canada to help build the railroads?
   a. Japanese
   b. Germans
   c. Chinese
   d. Mexicans

_____ **5.** The central government of Canada is headed by a
   a. president.
   b. premier.
   c. king.
   d. prime minister.

_____ **6.** How did the Canadians create a nation running from sea to sea?
   a. with airplanes
   b. with railroads
   c. with ships
   d. with automobiles

_____ **7.** The administrative divisions of Canada are known as
   a. Métis.
   b. provinces.
   c. agencies.
   d. dominions.

_____ **8.** The largest city in Canada is now
   a. Toronto.
   b. Quebec.
   c. Alberta.
   d. Saskatchewan.

Post-Reading Quick Check • After you have finished reading the section, in the space provided, explain why it can be said that the culture of Canada is a mosaic.

_____

_____

_____

_____

_____

_____

Canada

Reading the Section • As you read the section, match each place in the left column with its description in the right column. Write the letter of the correct description in the space provided.

_____ **1.** Quebec

_____ **2.** Nova Scotia

_____ **3.** Ontario

_____ **4.** Toronto

_____ **5.** Ottawa

_____ **6.** Alberta

_____ **7.** Edmonton

_____ **8.** British Columbia

_____ **9.** Vancouver

_____ **10.** Nunavut

a. Province in which farming and oil and natural gas production are important

b. Province near the ocean that occupies a peninsula

c. Canada's westernmost province

d. City between Toronto and Montreal where many people speak both English and French

e. Multicultural city with large Chinese and Indian populations

f. Canada's most populous province

g. New territory that was created for the native Inuit who live there

h. Province whose residents want to be independent from Canada

i. Largest city in Canada

j. Alberta's capital and largest city

Post-Reading Quick check • After you have finished reading the section, in the space provided, identify Canada's three Maritime Provinces, two Heartland Provinces, and three Prairie Provinces.

1. Maritime Provinces: _____ _____ _____

2. Heartland Provinces: _____ _____

3. Prairie Provinces: _____ _____ _____

 **Mexico**

SECTION 1

Reading the Section • As you read the section, complete each sentence below by writing the appropriate word or phrase in the space provided.

1. The country of Mexico is almost three times the size of _____.

2. Mexico has a long _____ Ocean coast and a shorter coast on the

Gulf of _____.

3. The Sierra Madre _____ to the east and the Sierra Madre

_____ to the west form the edges of the Mexican Plateau.

4. The capital of Mexico is _____.

5. In the Yucatán Peninsula, erosion has resulted in a large number of caves and

_____, or steep-sided depressions.

6. Mexico extends from the middle latitudes into the _____.

7. _____ is the rainy season in the forested plains along Mexico's southeastern coast.

8. Freezing temperatures sometimes reach as far south as _____.

9. Mexico's most important mineral resource is _____.

10. _____ is the most valuable part of Mexico's mining industry.

Post-Reading Quick Check • After you have finished reading the section, in the space provided, identify the four climate types found in Mexico.

1. Climate type: _____

2. Climate type: _____

3. Climate type: _____

4. Climate type: _____

CHAPTER 8 Mexico

Reading the Section • As you read the section, examine the descriptions of Mexican cultures below. For each description, place a check mark in the box next to the culture being described.

1. Made complex astronomical calculations and had a detailed calendar
❑ Olmec   ❑ Maya   ❑ Aztec   ❑ Colonial Mexico

2. Brought enslaved Africans to the region as a source of labor
❑ Olmec   ❑ Maya   ❑ Aztec   ❑ Colonial Mexico

3. Lived along the humid southern coast of the Gulf of Mexico around 1500 B.C.
❑ Olmec   ❑ Maya   ❑ Aztec   ❑ Colonial Mexico

4. Built their capital, Tenochtitlán, on an island in a lake in the Valley of Mexico
❑ Olmec   ❑ Maya   ❑ Aztec   ❑ Colonial Mexico

5. Collapsed sometime after A.D. 800, but descendants still live in the region today
❑ Olmec   ❑ Maya   ❑ Aztec   ❑ Colonial Mexico

6. Conquered by Hernán Cortés and his conquistadores in 1521
❑ Olmec   ❑ Maya   ❑ Aztec   ❑ Colonial Mexico

7. Caused the death of many Indians due to disease and overwork
❑ Olmec   ❑ Maya   ❑ Aztec   ❑ Colonial Mexico

8. Built temples, pyramids, and huge statues, and traded jade and obsidian
❑ Olmec   ❑ Maya   ❑ Aztec   ❑ Colonial Mexico

Post-Reading Quick Check • After you have finished reading the section, in the space provided, briefly describe the history of Mexico from the fight for independence to the Mexican Revolution.

_____

_____

_____

_____

_____

_____

Mexico

Reading the Section • As you read the section, complete the chart below by providing information about each of Mexico's six culture regions.

Greater Mexico City
Description:

Central Interior
Description:

Oil Coast
Description:

Southern Mexico
Description:

Northern Mexico
Description:

The Yucatán
Description:

Post-Reading Quick Check • After you have finished reading the section, in the space provided, identify three economic problems faced by Mexico in recent decades.

1. Economic problem: _____

2. Economic problem: _____

3. Economic problem: _____

Name _____ Class _____ Date _____

Reading the Section • As you read the section, circle the boldface word or phrase that *best* completes each statement below.

1. Central America and the Caribbean Sea include **60 / 20** countries and a number of island territories.

2. **Mountains / Deserts** separate the Caribbean and Pacific coastal plains.

3. The Caribbean islands form what is known as an archipelago—a large group of **islands / peninsulas**.

4. The four large islands of the **Greater / Lesser** Antilles are Cuba, Jamaica, Puerto Rico, and Hispaniola.

5. The islands in the **Greater / Lesser** Antilles stretch from the Virgin Islands to Trinidad and Tobago.

6. The Bahamas are located **west / east** of Florida.

7. The tectonic activity in the region is caused by **colliding / sliding** tectonic plates.

8. The Pacific coast has a warm and sunny **steppe / tropical savanna** climate.

9. In the islands, **winters / summers** usually are drier than **winters / summers**.

10. **Tornadoes / Hurricanes** are common in the region of Central America and the Caribbean islands.

Post-Reading Quick Check • After you have finished reading the section, in the space provided, discuss the major resources of Central America and the Caribbean islands.

_____

_____

_____

_____

_____

_____

_____

_____

CHAPTER 9

Central America and the Caribbean Islands

SECTION 2

Reading the Section • As you read the section, use the space provided to identify the country described by each statement. Choose your answers from the list below. Some answers will be used more than once.

Guatemala   Belize   Honduras   El Salvador
          Nicaragua   Costa Rica   Panama

_____ **1.** Volcanic ash has made this country's soils the most fertile in the region.

_____ **2.** Free elections here in 1990 ended the rule of the Sandinistas.

_____ **3.** This country is the most populous country in Central America.

_____ **4.** This country's canal was under U.S. control from 1914 to 1999.

_____ **5.** Fruit is an important export here.

_____ **6.** This country has a long history of stable, democratic government.

_____ **7.** Most people in this country live in poverty, which contributed to a long civil war in the 1980s.

_____ **8.** Many people here speak Maya languages.

_____ **9.** San José, this country's capital, is located in the central highlands.

_____ **10.** This country has the smallest population in Central America.

_____ **11.** This country is the largest Central American country.

_____ **12.** Transportation is difficult in the rugged terrain of this country.

Post-Reading Quick Check • After you have finished reading the section, in the space provided, number the following events in the order in which they occurred.

_____ **1.** Panama becomes independent.

_____ **2.** The Maya build large cities with pyramids and temples.

_____ **3.** British Honduras gains independence as Belize.

_____ **4.** Honduras, Guatemala, Nicaragua, El Salvador, and Costa Rica win independence.

Central America and the Caribbean Islands

SECTION 3

Reading the Section • As you read the section, match each place in the right column with its description in the left column. Write the letter of the correct place in the space provided. Answers may be used more than once.

_____ **1.** Occupies the mountainous western third of Hispaniola

_____ **2.** Commonwealth of the United States

_____ **3.** Largest country in the Caribbean

_____ **4.** Its capital, Santo Domingo, was the first permanent European settlement in the Western Hemisphere

_____ **5.** Its citizens are U.S. citizens

_____ **6.** Its capital and center of industry is Port-au-Prince

_____ **7.** Occupies the eastern part of Hispaniola

_____ **8.** Has had a communist government since Fidel Castro seized power in 1959

_____ **9.** Its economy is much more developed than the economies of other Caribbean islands

_____ **10.** Most of its farmlands are organized into cooperatives and government-owned sugarcane plantations

a. Cuba

b. Haiti

c. Dominican Republic

d. Puerto Rico

Post-Reading Quick Check • After you have finished reading the section, in the space provided, identify three types of Caribbean music and their islands of origin.

1. Type of music: _____ Island: _____

2. Type of music: _____ Island: _____

3. Type of music: _____ Island: _____

Caribbean South America

SECTION 1

Reading the Section • As you read the section, complete the following outline by supplying the main idea and the missing subtopics and supporting details.

Physical Geography

Main Idea: _____

Topic I: The region includes Colombia, Venezuela, Guyana, Suriname, and French Guiana.

Detail A: In the west, the Andes form a three-pronged cordillera.

Detail B: _____

Detail C: Between the two upland areas are the vast plains of the Orinoco River basin.

Detail D: _____

Topic II: _____

Detail A: Sugarcane and bananas grow in the tierra caliente, hot and humid lower elevations.

Detail B: _____

Detail C: Potatoes and wheat grow in the tierra fría, which has forests and grasslands.

Detail D: _____

Detail E: _____

Topic III: Caribbean South America has important resources.

Detail A: _____

Detail B: The region has mineral deposits, timber, fish and shrimp, and hydroelectric power.

Post-Reading Quick Check • After you have finished reading the section, in the space provided, match the Spanish terms with their English translations.

____ **1.** tierra caliente   ____ **3.** tierra fría   **a.** temperate country   **c.** hot country
____ **2.** tierra templada   ____ **4.** tierra helada   **b.** frozen country   **d.** cold country

People, Places, and Change

Name _____ Class _____ Date _____

Caribbean South America

CHAPTER 10

SECTION 2

Reading the Section • As you read the section, examine the riddles below. Solve each riddle by writing the correct word or words in the space provided.

_____ **1.** "We had a well-developed civilization in western Colombia. Our gold objects were among the finest to be found in ancient America. Who are we?"

_____ **2.** "I am the legend of 'the Golden One,' which describes a marvelous, rich land. What legend am I?"

_____ **3.** "When we arrived on the Caribbean coast around 1500 we were looking for a route to the Pacific Ocean. Instead, we conquered the Chibcha and seized their treasure. Who are we?"

_____ **4.** "I am the republic that was formed after the people of Central and South America won independence from Spain in the late 1700s. What republic am I?"

_____ **5.** "I am the national capital of present-day Colombia. You can find me high up in the eastern Andes. Come visit me! What city am I?"

_____ **6.** "Colombians grow me in their rich soil, and I am world-famous. Only Brazil produces more of me than Colombia. What am I?"

_____ **7.** "I am a tropical plant that is an important food crop for Colombians. I have starchy roots. What am I?"

_____ **8.** "I am a type of ringtoss game that many Colombians like to play. What am I?"

Post-Reading Quick Check • After you have finished reading the section, in the space provided, supply the requested information about Colombia below.

1. Leading export: _____

2. Two problems: _____

3. Two sports: _____

4. Main religion: _____

People, Places, and Change

Caribbean South America

Reading the Section • As you read the section, examine each of the pairs of statements below. Circle the letter of the statement in each pair that is true.

1. **a.** By the early 1500s the Spanish were exploring the area of Venezuela.
   b. By the early 1500s the Vikings were exploring the area of Venezuela.

2. **a.** Early settlers in Venezuela grew indigo, which became their main food crop.
   b. Early settlers in Venezuela grew indigo, which is used to make a deep blue dye.

3. **a.** Because he led wars of independence, Simón Bolívar is a hero in South America.
   b. Because he led wars of independence, Christopher Columbus is a hero in South America.

4. **a.** The term *caudillos* refers to Venezuela's major exports.
   b. The term *caudillos* refers to Venezuela's military leaders.

5. **a.** By the 1970s Venezuela was earning huge sums of money from diamonds.
   b. By the 1970s Venezuela was earning huge sums of money from oil.

6. **a.** About 85 percent of Venezuelans live in cities and towns.
   b. About 15 percent of Venezuelans live in cities and towns.

7. **a.** The national capital of Venezuela is Caracas.
   b. The national capital of Venezuela is Maracaibo.

8. **a.** Venezuela's economy is based on oil production.
   b. Venezuela's economy is based on agriculture.

9. **a.** Llaneros are Venezuela's most unusual animals.
   b. Llaneros are Venezuelan cowboys.

10. **a.** The official language of Venezuela is Spanish.
    b. The official language of Venezuela is Portuguese.

Post-Reading Quick Check • After you have finished reading the section, in the space provided, discuss cultural life in Venezuela.

_____

_____

_____

_____

_____

Caribbean South America

Reading the Section • All of the following statements refer to the Guianas. As you read the section, use the space provided to indicate whether each statement below refers to Guyana, Suriname, or French Guiana.

_____ **1.** Formerly known as Dutch Guiana, it gained independence in 1975.

_____ **2.** One of its major exports is aluminum.

_____ **3.** It remains part of France.

_____ **4.** Nearly half of its people live in Paramaribo, the capital.

_____ **5.** Its name is a South American Indian word that means "land of waters."

_____ **6.** It has a status in France similar to a state in the United States.

_____ **7.** It depends heavily on imports for food and energy.

_____ **8.** Its major mineral resource is bauxite.

_____ **9.** Its capital is called Cayenne.

_____ **10.** More than a third of its population lives in Georgetown, the capital.

Post-Reading Quick Check • After you have finished reading the section, in the space provided, explain why European colonists brought indentured servants to the Guianas.

_____

_____

_____

_____

_____

_____

_____

Atlantic South America

SECTION 1

Reading the Section • As you read the section, complete the chart below by providing information about the physical features of Atlantic South America in each of the categories shown.

Plains and Plateaus
Information:

Mountains
Information:

River Systems
Information:

The Rain Forest
Information:

Climates
Information:

Resources
Information:

Post-Reading Quick Check • After you have finished reading the section, in the space provided, identify the four countries that make up Atlantic South America.

1. Country: _____

2. Country: _____

3. Country: _____

4. Country: _____

People, Places, and Change

Guided Reading Activities **33**

Atlantic South America

CHAPTER 11

Reading the Section • As you read the section, complete each sentence below by writing the appropriate word or phrase in the space provided.

1. Brazilian Indians are descended from people who probably came from

_____.

2. After 1500, _____ settlers began to move into the region of Brazil.

3. In the late 1800s southeastern Brazil became a major producer of _____, which spurred the growth of the city of São Paulo.

4. In terms of religion, Brazil is the world's largest _____ country.

5. During Carnival, Brazilians dance the _____, which was adapted from an African dance.

6. Brazil is the _____-largest country in the world in terms of land area and population.

7. The _____ region covers much of northern and western Brazil.

8. Brazil's poorest region is the _____ region.

9. The _____ region is Brazil's richest region.

10. _____, the capital of Brazil, is located in the interior region.

Post-Reading Quick Check • After you have finished reading the section, in the space provided, discuss the history of Brazil's government.

_____

_____

_____

_____

_____

_____

CHAPTER 11

Atlantic South America

SECTION 3

Reading the Section • As you read the section, use each clue below to write the correct term in the space provided. Then unscramble the boxed letters to identify the focus of Section 3.

1. Meaning of *Argentina*: "_ ☐ __ __ __ __ _ ☐ __ __ __ __ __"

2. Spanish landowners had right to labor of Indians: ☐ __ __ __ __ __ __ __ __ __

3. Term for Argentine cowboys: ☐ __ __ __ __ __ __

4. Islands Argentina fought over in 1983: __ __ __ __ __ __ __ ☐ __

5. Major religion in Argentina: __ __ __ __ __   __ __ ☐ __ __ __ __ __

6. Sausages and steaks served on a small grill: __ ☐ __ __ __ __ __ __ __

7. Capital of Argentina: __ __ __ ☐ __ __   __ __ __ __ __

8. Trade organization that promotes economic cooperation among its members in

southern and eastern South America: __ __ __ __ __ __ __ ☐

The focus of Section 3 is: ☐ ☐ ☐ ☐ ☐ ☐ ☐ ☐ ☐

Post-Reading Quick Check • After you have finished reading the section, in the space provided, write the following information about Argentina.

1. Important agricultural region during colonial era: _____

2. Year Argentina gained independence: _____

3. Official language: _____

4. Argentine dance: _____

5. Argentine dance: _____

6. Most developed agricultural region: _____

7. Percentage of labor force in agriculture: _____

8. Government: _____

Atlantic South America

CHAPTER 11

SECTION 4

Reading the Section • As you read the section, consider each of the statements listed below. In the space provided, write "U" if the statement refers to Uruguay, "P" if the statement refers to Paraguay, or "B" if the statement refers to both countries.

_____ **1.** More than 90 percent of the people live in urban areas.

_____ **2.** It lies along the Río de la Plata, the major waterway of southern South America.

_____ **3.** It shares borders with Bolivia, Brazil, and Argentina.

_____ **4.** The main religion is Roman Catholicism.

_____ **5.** It declared independence from Spain in 1811.

_____ **6.** Its capital, Montevideo, is the country's business and government center.

_____ **7.** About 95 percent of the people are mestizos.

_____ **8.** Spanish is the official language.

_____ **9.** Its capital and largest city is Asunción.

_____ **10.** About 88 percent of the population is of European descent.

Post-Reading Quick Check • After you have finished reading the section, in the space provided, discuss the energy resources of both Uruguay and Paraguay.

1. Uruguay: _____

_____

_____

_____

2. Paraguay: _____

_____

_____

_____

Pacific South America

CHAPTER 12

SECTION 1

Reading the Section • As you read the section, examine each of the pairs of statements below. Circle the letter of the statement in each pair that is true.

1. **a.** The country of Bolivia lies on the equator.
   b. The country of Ecuador lies on the equator.

2. **a.** The Andes run through all four of the region's countries.
   b. The Himalayas run through all four of the region's countries.

3. **a.** The Strait of Magellan links the Atlantic Ocean and the Pacific Ocean.
   b. The Strait of Magellan links the Gulf of Mexico and the Atlantic Ocean.

4. **a.** The Altiplano lies between southern Chile and Ecuador.
   b. The Altiplano lies between southern Peru and Bolivia.

5. **a.** Eastern Ecuador and Peru and northern Bolivia share a steppe climate.
   b. Eastern Ecuador and Peru and northern Bolivia share a humid tropical climate.

6. **a.** South Americans call the tropical rain forests in the eastern region *selvas*.
   b. South Americans call the low-lying clouds in the eastern region *selvas*.

7. **a.** Rain is rare in the Atacama Desert, but fog and low clouds are common.
   b. Fog and low clouds are rare in the Atacama Desert, but rain is common.

8. **a.** Coastal Peru is one of the cloudiest and driest places on Earth.
   b. Coastal Chile is one of the cloudiest and driest places on Earth.

9. **a.** Every two to seven years, the dry Pacific coast is affected by an ocean and weather pattern called the Peru Current.
   b. Every two to seven years, the dry Pacific coast is affected by an ocean and weather pattern called El Niño.

10. **a.** Chile has very large deposits of tin.
    b. Bolivia has very large deposits of tin.

Post-Reading Quick Check • After you have finished reading the section, in the space provided, explain what causes El Niño conditions.

_____

_____

_____

_____

Pacific South America

Reading the Section • As you read the section, number the following events in the order in which they occurred.

_____ **1.** Francisco Pizarro and his Spanish soldiers defeat the Inca.

_____ **2.** Chileans reject military dictatorship and create a democratic government.

_____ **3.** The country of Ecuador achieves independence from Spain.

_____ **4.** Peru's first advanced civilization reaches its height.

_____ **5.** The Inca Empire covers about a million square miles and houses 12 million people.

_____ **6.** Peru becomes an independent country two years after Ecuador does so.

_____ **7.** The Inca emperor dies, and a struggle begins between two of his sons over who will take his place.

_____ **8.** Chile becomes an independent country.

_____ **9.** Tupac Amarú II leads Indians in a revolt against their Spanish rulers.

_____ **10.** Bolivia becomes the last country in Pacific South America to achieve independence.

Post-Reading Quick Check • After you have finished reading the section, in the space provided, list six important achievements of the Inca civilization.

1. Achievement: _____

2. Achievement: _____

3. Achievement: _____

4. Achievement: _____

5. Achievement: _____

6. Achievement: _____

Pacific South America

CHAPTER 12

SECTION 3

Reading the Section • As you read the section, examine the descriptions of Pacific South America below. For each description, place a check mark in the box next to the country in Pacific South America being described.

1. The capital of this country is the highest capital in the world.
   ❑ Ecuador   ❑ Bolivia   ❑ Peru   ❑ Chile

2. In the 1980s this country ended the rule of a junta and became a democracy.
   ❑ Ecuador   ❑ Bolivia   ❑ Peru   ❑ Chile

3. Its largest city, Guayaquil, is located in the coastal lowland.
   ❑ Ecuador   ❑ Bolivia   ❑ Peru   ❑ Chile

4. Hydroelectric projects on this country's coastal rivers provide energy.
   ❑ Ecuador   ❑ Bolivia   ❑ Peru   ❑ Chile

5. This country has the highest percentage of Indians of any South American country.
   ❑ Ecuador   ❑ Bolivia   ❑ Peru   ❑ Chile

6. Many people have suggested that this country become the next country to join NAFTA.
   ❑ Ecuador   ❑ Bolivia   ❑ Peru   ❑ Chile

7. Its rich oil deposits are essential to its economy.
   ❑ Ecuador   ❑ Bolivia   ❑ Peru   ❑ Chile

8. This country houses Cuzco and Machu Picchu, stone structures dating from the Inca period.
   ❑ Ecuador   ❑ Bolivia   ❑ Peru   ❑ Chile

Post-Reading Quick Check • After you have finished reading the section, in the space provided, identify the capital city of each country listed below.

1. Ecuador: _____

2. Bolivia: _____

3. Peru: _____

4. Chile: _____

Southern Europe

EASTERN CHAPTER 6

SECTION 1

Reading the Section • All of the following statements refer to the physical geography of countries in southern Europe. As you read the section, use the space provided to indicate whether each statement below refers to Portugal, Spain, Italy, or Greece.

_____ **1.** The main landmass of this country extends into the Aegean Sea.

_____ **2.** The Po is this country's largest river.

_____ **3.** Along with Spain, this country occupies the Iberian Peninsula.

_____ **4.** The Balearic Islands are part of this country.

_____ **5.** Lisbon, the capital of this country, is an important Atlantic port.

_____ **6.** This country's peninsula is shaped like a boot.

_____ **7.** The people in this country mine bauxite, chromium, lead, and zinc.

_____ **8.** This country is mountainous and includes more than 2,000 islands.

_____ **9.** In this country semiarid climates are found in pockets.

_____ **10.** This country's far north region includes the southern Alps.

Post-Reading Quick Check • After you have finished reading the section, in the space provided, explain why visitors have long been attracted to southern Europe.

_____

_____

_____

_____

_____

_____

_____

_____

_____

Southern Europe

EASTERN CHAPTER 6

SECTION 2

Reading the Section • As you read the section, answer each of the following questions in the space provided.

1. What made up a Greek *polis*? _____

2. Which Greek city-state was the first known democracy? _____

3. Whose empire combined Greek culture with Asian and African influences? _____

4. What name was given to the eastern half of the Roman Empire? _____

5. What is the leading form of Christianity in Greece? _____

6. Which people conquered Constantinople in 1453? _____

7. What Greek industry employs the most people? _____

8. Approximately what percentage of Greeks live in rural areas? _____

9. What is both the capital of Greece and its largest city? _____

10. What is the name of Athens's seaport? _____

Post-Reading Quick Check • After you have finished reading the section, in the space provided, discuss the art forms for which Greece is famous.

_____

_____

_____

_____

_____

_____

_____

_____

_____

Southern Europe

EASTERN CHAPTER 6

SECTION 3

Reading the Section • As you read the section, examine the riddles below. Solve each riddle by writing the correct word or name in the space provided.

_____ 1. "Around 750 B.C. we established the city of Rome on the Tiber River, a beautiful location for a city. Who are we?"

_____ 2. "Those Romans were such clever engineers—they built us to transport water through the city. What are we?"

_____ 3. "Even though my western relative, with its capital in Rome, fell in A.D. 476, I lasted until 1453. What am I?"

_____ 4. "I am the Roman province in which Christianity got its start. Which province am I?"

_____ 5. "As the bishop of Rome, I am the head of the Roman Catholic Church. By what title am I known?"

_____ 6. "I don't like to brag, but in addition to painting the _Mona Lisa_, I was a sculptor, engineer, architect, and scientist. Who am I?"

_____ 7. "I was such a great explorer that North America and South America were named after me. Who am I?"

_____ 8. "Even though I originated in Naples, I am a food that Americans love to eat. Have you tried my pepperoni? What food am I?"

_____ 9. "I am Italy's most valuable crop, and I am grown throughout the country. Italians use me to make wine. Which crop am I?"

_____ 10. "I am the city in Italy famous for my romantic canals and beautiful buildings. Come visit me! Which city am I?"

Post-Reading Quick Check • After you have finished reading the section, in the space provided, identify the foods eaten by people in southern Italy and northern Italy.

1. Southern Italy: _____

_____

2. Northern Italy: _____

_____

People, Places, and Change

CHAPTER 13 Southern Europe

SECTION 4

EASTERN CHAPTER 6

Reading the Section • Each of the following sentences contains an underlined word or name that makes the sentence incorrect. As you read the section, use the space provided to write the word or phrase that makes the sentence correct.

_____ **1.** The Iberian Peninsula is occupied by two countries—<u>Italy</u> and Portugal.

_____ **2.** Around 200 B.C. Iberia became part of the Roman Empire and adopted the <u>French</u> language.

_____ **3.** The Muslim North Africans, or <u>Granadians</u>, conquered most of the Iberian Peninsula in the A.D. 700s.

_____ **4.** King Ferdinand and Queen <u>Elizabeth</u> sponsored the voyage of Christopher Columbus to the Americas.

_____ **5.** In 1588 Philip II, the king of Spain and Portugal, sent a huge armada to invade <u>France</u>, but was defeated.

_____ **6.** Today Spain is a democracy, with a national assembly and a <u>president</u>.

_____ **7.** Castilian is the most widely understood <u>Portuguese</u> dialect.

_____ **8.** Both Spain and Portugal are strongly <u>Protestant</u>.

_____ **9.** <u>Barcelona</u> is Portugal's capital and largest city.

_____ **10.** Spain's capital and largest city is <u>Lisbon</u>.

Post-Reading Quick Check • After you have finished reading the section, in the space provided, discuss the government of Spain during the 1900s.

_____

_____

_____

_____

_____

West-Central Europe

EASTERN CHAPTER 7

Reading the Section • As you read the section, complete the following outline by supplying the main idea and the missing subtopics and supporting details.

Physical Geography

Main Idea: _____

Topic I: West-central Europe includes France, Germany, the Benelux countries, and the Alpine countries.

    Detail A: The landforms of west-central Europe are arranged like a fan.

    Detail B: _____

    Detail C: West-central Europe contains the Alps, Europe's highest mountain range.

Topic II: _____

    Detail A: A marine west coast climate makes west-central Europe very pleasant.

    Detail B: _____

Topic III: West-central Europe has important resources.

    Detail A: The region has some of the most productive fields in the world.

    Detail B: _____

    Detail C: _____

Post-Reading Quick Check • After you have finished reading the section, in the space provided, identify nine major rivers in west-central Europe.

1. River: _____   **6.** River: _____

2. River: _____   **7.** River: _____

3. River: _____   **8.** River: _____

4. River: _____   **9.** River: _____

5. River: _____

West-Central Europe

SECTION 2

EASTERN CHAPTER 7

Reading the Section • As you read the section, write the letter of the *best* choice in the space provided.

_____ **1.** In what year did the French Revolution began?
   a. 1914   **c.** 1789
   b. 1812   **d.** 1776

_____ **2.** Which group took over much of Gaul after the Roman Empire collapsed?
   a. the Slavs
   b. the Vandals
   c. the Vikings
   d. the Franks

_____ **3.** After the fall of the Roman Empire, much of western Europe was ruled by
   a. Renoir.
   b. Charlemagne.
   c. Napoléon Bonaparte.
   d. Gauguin.

_____ **4.** The Hundred Years' War began when the king of England tried to claim the throne
   a. of Africa.
   b. of France.
   c. of Finland.
   d. of Italy.

_____ **5.** The largest city in France is
   a. Marseille.
   b. Lyon.
   c. Nice.
   d. Paris.

_____ **6.** NATO is
   a. an acronym for the countries that fought in World War II.
   b. France's most important export.
   c. the new currency of Europe.
   d. a military alliance created to defend Western Europe.

_____ **7.** About 90 percent of French people are
   a. Jewish.
   b. Protestant.
   c. Roman Catholic.
   d. Basque.

_____ **8.** Present-day Normandy was originally settled by Normans from
   a. Scandinavia.
   b. France.
   c. Russia.
   d. South America.

Post-Reading Quick Check • After you have finished reading the section, in the space provided, discuss when, why, and how the French people celebrate Bastille Day.

_____

_____

_____

_____

_____

West-Central Europe

EASTERN CHAPTER 7

Reading the Section • As you read the section, examine each of the pairs of statements below. Circle the letter of the statement in each pair that is true.

1. **a.** When the Roman Empire fell, the Franks became the most important tribe in Germany.
   b. When the Roman Empire fell, the Normans became the most important tribe in Germany.

2. **a.** The effort to reform Christianity during the 1500s is known as the Reformation.
   b. The effort to reform Christianity during the 1500s is known as the Renaissance.

3. **a.** During the late 1800s Berlin led the creation of a united Germany.
   b. During the late 1800s Prussia led the creation of a united Germany.

4. **a.** Germany and its allies won World War I, but were defeated in World War II.
   b. Germany and its allies were defeated in World War I and World War II.

5. **a.** In 1990 Germany was split into two countries—East Germany and West Germany.
   b. In 1990 East Germany and West Germany were reunited into one country.

6. **a.** Nine out of 10 inhabitants of Germany are ethnic Germans.
   b. Only 2 out of 10 inhabitants of Germany are ethnic Germans.

7. **a.** The major German festival season is the 4th of July.
   b. The major German festival season is Christmas.

8. **a.** Germany is one of the world's leading industrial countries.
   b. More people in Germany work in agriculture than in any other occupation.

9. **a.** Germany's capital city is Munich, which was isolated after World War II.
   b. Germany's capital city is Berlin, which was isolated after World War II.

Post-Reading Quick Check • After you have finished reading the section, in the space provided, describe the government of Germany since reunification.

_____

_____

_____

_____

CHAPTER 14

West-Central Europe

EASTERN CHAPTER 7

SECTION 4

Reading the Section • As you read the section, complete each
sentence below by writing the appropriate word, phrase, or place
in the space provided.

1. The Netherlands is sometimes called _____, and its people are
known as the Dutch.

2. The Protestants of the Netherlands won their freedom from _____
rule in the 1570s.

3. Belgium has been ruled at times by _____ and by the Netherlands.

4. Many of the major battles that were fought during World War I were fought in

_____ .

5. Each of the three Benelux countries today is ruled by a(n) _____
and a monarch.

6. _____ is the official language of the Netherlands.

7. The foods of the Benelux region emphasize _____ products and
sausage.

8. The Dutch brought their "Koekjes" to the Americas; Americans now call these little

cakes _____ .

9. The Netherlands is famous for its flowers, particularly its _____ .

10. Luxembourg today earns much of its income from _____ .

Post-Reading Quick Check • After you have finished reading
the section, in the space provided, discuss the religious preferences
of the people living in the Benelux countries.

_____

_____

_____

_____

_____

West-Central Europe

EASTERN CHAPTER 7

SECTION 5

Reading the Section • As you read the section, consider each of the statements listed below. In the space provided, write "S" if the statement refers to Switzerland or write "A" if the statement refers to Austria.

_____ **1.** It gained its independence in the 1600s, after its cantons gradually broke away from the Holy Roman Empire.

_____ **2.** This country is overwhelmingly Roman Catholic.

_____ **3.** The capital city of this country is Bern.

_____ **4.** This country is famous for its watches, optical instruments, and other machinery.

_____ **5.** It has remained strictly neutral in the European wars of the past two centuries.

_____ **6.** It was the home of the Habsburgs, a powerful family of German nobles.

_____ **7.** The composer Mozart wrote symphonies and operas here.

_____ **8.** Vienna is its commercial and industrial center.

_____ **9.** Today this country is a confederation of 26 cantons.

_____ **10.** This country's people are almost overwhelmingly German-speaking.

Post-Reading Quick Check • After you have finished reading the section, in the space provided, identify when the Red Cross Organization was founded, where it was founded, and why it was founded.

1. When: _____

2. Where: _____

3. Why: _____

_____

Northern Europe

EASTERN CHAPTER 8

CHAPTER 15

SECTION 1

Reading the Section • As you read the section, match each term or place in the left column with its description in the right column. Write the letter of the correct description in the space provided.

_____ **1.** Greenland

_____ **2.** Denmark

_____ **3.** Iceland

_____ **4.** fjord

_____ **5.** loch

_____ **6.** Shannon

_____ **7.** North Sea

_____ **8.** Baltic Sea

_____ **9.** Russia

_____ **10.** North Atlantic Drift

a. Country that lies on the Jutland Peninsula and nearby islands

b. Warm ocean current in northern Europe

c. Longest river in the British Isles

d. Narrow, deep inlet of the sea between high, rocky cliffs

e. Ice-free body of water that is especially important for trade and fishing

f. One country from which northern Europe imports oil and natural gas

g. Country that is mountainous and volcanic and that has more than 100 glaciers

h. Sea that freezes over during the winter months

i. World's largest island

j. Scottish term for a lake in a valley carved by a glacier long ago

Post-Reading Quick Check • After you have finished reading the section, in the space provided, discuss the forests and soil of northern Europe.

_____

_____

_____

_____

_____

_____

_____

Northern Europe

EASTERN CHAPTER 8

SECTION 2

Reading the Section • As you read the section, circle the boldface word or phrase that *best* completes each statement below.

1. The last group of people to arrive in Britain were the **Normans / Saxons**.

2. By 1900 the British Empire covered nearly **one half / one fourth** of the world's land area.

3. The United Kingdom's early industries were powered by **electricity / coal**.

4. In the United Kingdom the **parliament / king or queen** makes the country's laws.

5. The United Kingdom today is home to nearly **60 / 260** million people.

6. **Welsh / English** is the official language of the United Kingdom.

7. The British people enjoy fish and chips, or fried fish and **tomatoes / potatoes**.

8. In June of every year the British celebrate the **prime minister's / queen's** birthday.

9. The United Kingdom's capital and largest city is **London / Belfast**.

10. **Race / Religion** has been the cause of much violence in Northern Ireland.

Post-Reading Quick Check • After you have finished reading the section, in the space provided, discuss agriculture in the United Kingdom.

_____

_____

_____

_____

_____

_____

_____

_____

Northern Europe

CHAPTER 15

EASTERN CHAPTER 8

Reading the Section • As you read the section, complete the graphic organizer by supplying information about the Republic of Ireland in each of the categories shown.

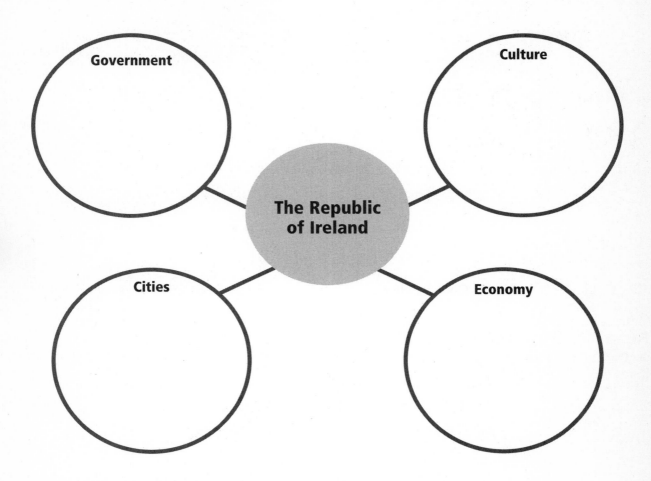

Government

Culture

The Republic of Ireland

Cities

Economy

Post-Reading Quick Check • After you have finished reading the section, in the space provided, provide two reasons why many Irish people left Ireland in the 1800s.

1. Reason: _____

_____

2. Reason: _____

_____

Northern Europe

CHAPTER 15

SECTION 4

EASTERN CHAPTER 8

Reading the Section • As you read the section, use the space provided to identify the country described by each statement. Choose your answers from the list below. Some answers will be used more than once.

Norway   Sweden   Denmark   Greenland

Iceland   Finland   Lapland

_____ **1.** This country is Scandinavia's smallest and most densely populated country.

_____ **2.** Most of the people who live here are Inuit.

_____ **3.** This country's capital, Oslo, is one of its largest cities.

_____ **4.** The capital of this country and its largest city is Helsinki.

_____ **5.** This is Scandinavia's largest and most populous country.

_____ **6.** The Sami live in this cultural region.

_____ **7.** Fjords shelter this country's harbors and its shipping fleets.

_____ **8.** This island country's geysers heat its homes and greenhouses.

_____ **9.** This country's North Sea oil fields probably will run dry soon.

_____ **10.** Nearly a fourth of this country's population lives in and around Copenhagen, the capital.

_____ **11.** This country is the easternmost country in the region.

_____ **12.** This country remained neutral during World Wars I and II.

Post-Reading Quick Check • After you have finished reading the section, in the space provided, identify three similarities found among the countries of Scandinavia.

1. Similarity: _____

2. Similarity: _____

3. Similarity: _____

Eastern Europe

EASTERN CHAPTER 9

Reading the Section • As you read the section, answer each of the following questions in the space provided.

1. Which countries are the Baltic countries? _____

2. What river flows through the Great Hungarian Plain? _____

3. Into which peninsula do the Alps of central Europe extend? _____

4. In which country is the Transylvanian Alps? _____

5. What is Eastern Europe's most important river? _____

6. How many tributaries flow into the Danube? _____

7. What is the weather like in the eastern half of the region? _____

8. What is oil shale? _____

9. Which countries mine lignite? _____

10. What word is used to describe fossilized tree sap? _____

Post-Reading Quick Check • After you have finished reading the section, in the space provided, explain why Eastern Europe has pollution problems.

_____

_____

_____

_____

_____

_____

_____

_____

_____

_____

People, Places, and Change

Eastern Europe

CHAPTER 16

SECTION 2

EASTERN CHAPTER 9

Reading the Section • As you read the section, complete each
sentence below by writing the appropriate word, phrase, or place
in the space provided.

1. _____ refers to the collapse of communism in Czechoslovakia.

2. In the 1200s the Mongols under _____ invaded Hungary.

3. The Estonian language is closely related to _____ .

4. The population of _____ has the highest percentage of ethnic
minorities.

5. _____ is the southernmost Baltic country.

6. Northeastern Europe's largest and most populous country is _____ .

7. In 1993 _____ split into the Czech Republic and Slovakia.

8. Much of the industry in _____ is located in and around Prague.

9. Most of the people who live in _____ are Slovaks.

10. Much of the manufacturing in _____ is located in Budapest.

Post-Reading Quick Check • After you have finished reading
the section, in the space provided, identify the capital city of each
of the following countries.

1. Estonia: _____

2. Latvia: _____

3. Lithuania: _____

4. Czech Republic: _____

5. Slovakia: _____

6. Hungary: _____

Eastern Europe

SECTION 3

EASTERN CHAPTER 9

Reading the Section • As you read the section, examine the riddles below. Solve each riddle by writing the correct word or place in the space provided.

_____ **1.** "After we conquered Constantinople in 1453, we ruled over south-eastern Europe until the 1800s. Who are we?"

_____ **2.** "In 1817 I became southeastern Europe's first self-governing region. Ah, I remember it well. What region am I?"

_____ **3.** "World War I began when I declared war on Serbia for killing the heir to the Austro-Hungarian throne. What country am I?"

_____ **4.** "I am the capital of both Serbia and Yugoslavia. You can find me on the Danube River. Which city am I?"

_____ **5.** "My beautiful capital city of Sarajevo was heavily damaged in the early 1990s after I declared independence. What country am I?"

_____ **6.** "Even though I once was a part of Yugoslavia, I now try to avoid the ethnic fighting going on around there. What country am I?"

_____ **7.** "I am the part of the Romanian economy that employs more people than any other part of the economy. What am I?"

_____ **8.** "I am the type of economy in which consumers help determine what is to be produced. What type of economy am I?"

Post-Reading Quick Check • After you have finished reading the section, in the space provided, explain how the foods eaten in southeastern Europe reflect the surrounding areas.

_____

_____

_____

_____

_____

_____

Russia

EASTERN CHAPTER 10

Reading the Section • All of the following statements refer to Russia. As you read the section, use the space provided to indicate whether each statement below refers to Russia's physical features, Russia's climate and vegetation, or Russia's resources.

_____ **1.** Russia's vast taiga grows south of the tundra.

_____ **2.** Russia stretches 6,000 miles from Eastern Europe to the Bering Sea and Pacific Ocean.

_____ **3.** Russia has long been a major oil producer.

_____ **4.** Russia is a major producer of the world's diamonds.

_____ **5.** The Volga River, which flows through European Russia, is Europe's longest river.

_____ **6.** Cold arctic winds sweep across much of Russia in winter.

_____ **7.** Much of the forest west of the Urals has been cut down.

_____ **8.** The Ural Mountains divide Europe from Asia.

_____ **9.** A series of high mountain ranges runs through southern and eastern Siberia.

_____ **10.** Much of the Russian steppe is used to grow crops and graze livestock.

Post-Reading Quick Check • After you have finished reading the section, in the space provided, identify the four types of climate that are found in Russia.

1. Climate: _____

2. Climate: _____

3. Climate: _____

4. Climate: _____

Name _____ Class _____ Date _____

Russia

EASTERN CHAPTER 10

Reading the Section • As you read the section, examine the descriptions of Russia below. For each description, place a check mark in the box next to the period in Russia's history being described.

1. The Communists under Joseph Stalin take control of all industries and farms.
  ❏ Early Russia   ❏ Russian Empire   ❏ Soviet Union   ❏ Russia Today

2. Ivan IV crowns himself czar of all Russia.
  ❏ Early Russia   ❏ Russian Empire   ❏ Soviet Union   ❏ Russia Today

3. Russia is making the transition from communism to democracy and open markets.
  ❏ Early Russia   ❏ Russian Empire   ❏ Soviet Union   ❏ Russia Today

4. Vikings called the Rus help shape the first Russian state among the Slavs.
  ❏ Early Russia   ❏ Russian Empire   ❏ Soviet Union   ❏ Russia Today

5. Mikhail Gorbachev promotes open discussion and policies to help the economy.
  ❏ Early Russia   ❏ Russian Empire   ❏ Soviet Union   ❏ Russia Today

6. The Mongols conquer Kiev and add much of the region to their vast empire.
  ❏ Early Russia   ❏ Russian Empire   ❏ Soviet Union   ❏ Russia Today

7. The Russian Federation is governed by an elected president and a parliament.
  ❏ Early Russia   ❏ Russian Empire   ❏ Soviet Union   ❏ Russia Today

8. The Bolshevik Party, led by Vladimir I. Lenin, overthrows the Russian government.
  ❏ Early Russia   ❏ Russian Empire   ❏ Soviet Union   ❏ Russia Today

Post-Reading Quick Check • After you have finished reading the section, in the space provided, describe how the Mongols ruled Russia.

_____

_____

_____

_____

_____

_____

People, Places, and Change   **Guided Reading Activities 57**

CHAPTER
17

Russia

EASTERN CHAPTER 10

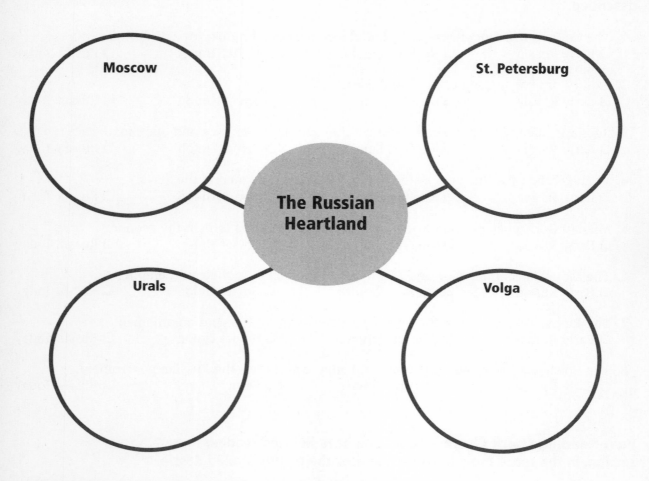

SECTION 3

Reading the Section • As you read the section, complete the graphic organizer by supplying information about the four major regions of the Russian heartland.

Moscow

St. Petersburg

The Russian Heartland

Urals

Volga

Post-Reading Quick Check • After you have finished reading the section, in the space provided, list four characteristics of European Russia that make it the Russian heartland.

1. Characteristic: _____

2. Characteristic: _____

3. Characteristic: _____

4. Characteristic: _____

Russia

EASTERN CHAPTER 10

SECTION 4

Reading the Section • As you read the section, provide the following information about Siberia.

1. Size: _____

2. Ocean to the north: _____

3. Countries to the south: _____

4. Meaning of *Siberia*: _____

5. Winters: _____

6. Major ethnic group: _____

7. Railroads: _____

8. Natural resources: _____

9. Most important industries: _____

10. Most important industrial region: _____

11. Largest city: _____

12. "Jewel of Siberia": _____

Post-Reading Quick Check • After you have finished reading the section, in the space provided, discuss how pollution threatens Lake Baikal and what is being done about this threat.

_____

_____

_____

_____

_____

_____

_____

Name _____ Class _____ Date _____

Russia

EASTERN CHAPTER 10

Reading the Section • As you read the section, examine each of the pairs of statements below. Circle the letter of the statement in each pair that is true.

1. a. Off the eastern coast of Siberia are the Sea of Okhotsk and the Sea of Japan.
 b. Off the western coast of Siberia are the Sea of Okhotsk and the Sea of Japan.

2. a. The Russian Far East has a less severe climate than the rest of Siberia.
 b. The Russian Far East has a more severe climate than the rest of Siberia.

3. a. The Russian Far East produces plentiful food for its people.
 b. The Russian Far East cannot produce enough food for its people.

4. a. Much of the Russian Far East remains forested.
 b. Most of the forests in the Russian Far East have been cut down.

5. a. An important energy resource in the Russian Far East is nuclear power.
 b. An important energy resource in the Russian Far East is geothermal energy.

6. a. The name *Vladivostok* means "Lovely City" in Russian.
 b. The name *Vladivostok* means "Lord of the East" in Russian.

7. a. People in Vladivostok use icebreakers to allow ships to use their waterways in winter.
 b. People in Vladivostok use icebreakers to bring water into their homes in winter.

8. a. Sakhalin and the Kurils sometimes experience earthquakes and volcanic eruptions.
 b. Sakhalin and the Kurils sometimes experience hurricanes and tornadoes.

Post-Reading Quick Check • After you have finished reading the section, in the space provided, discuss the argument over ownership of Sakhalin and the Kurils.

_____

_____

_____

_____

_____

_____

People, Places, and Change

Ukraine, Belarus, and the Caucasus

EASTERN CHAPTER 11

CHAPTER 18

SECTION 1

Reading the Section • As you read the section, complete each sentence below by writing the appropriate word, phrase, or place in the space provided.

1. The countries of Belarus and Ukraine border western _____.

2. Georgia, _____, and Azerbaijan lie in the Caucasus region.

3. The _____ Mountains run through part of western Ukraine.

4. Southern Crimea separates the _____ Sea from the Sea of Azov.

5. Mount _____, at 18,510 feet, is Europe's highest mountain.

6. One of Europe's rivers, the _____ River, flows south through Belarus and Ukraine.

7. _____ has replaced much of the region's original vegetation.

8. Ukraine is trying to preserve its natural environments and so has created several

   nature _____.

9. The northern two thirds of Ukraine and Belarus have a _____ climate.

10. The most important mineral resources in the region are Azerbaijan's large and valuable

    oil and _____ deposits.

Post-Reading Quick Check • After you have finished reading the section, in the space provided, discuss the region's farmlands.

_____

_____

_____

_____

_____

_____

People, Places, and Change

Ukraine, Belarus, and the Caucasus

EASTERN CHAPTER 11

SECTION 2

Reading the Section • As you read the section, consider each of the statements listed below. In the space provided, write "U" if the statement refers to Ukraine, "B" if the statement refers to Belarus, or "UB" if the statement refers to both countries.

_____ **1.** It is one of the world's top steel producers.

_____ **2.** The people use the Cyrillic alphabet.

_____ **3.** Its people are known as "white Russians."

_____ **4.** About 75 percent of the population is made up of ethnic Ukrainians.

_____ **5.** It has a large reserve of potash and is a world leader in the production of peat.

_____ **6.** It has some of the world's richest soil.

_____ **7.** Its capital, Minsk, is the administrative center of the Commonwealth of Independent States.

_____ **8.** It received radiation from the Chernobyl nuclear disaster.

_____ **9.** It is the world's largest producer of sugar beets.

_____ **10.** One of the country's main crops is flax, which is grown for its fiber and seed.

Post-Reading Quick Check • After you have finished reading the section, in the space provided, number the following events in the order in which they occurred.

_____ **1.** The Mongols conquer Ukraine and destroy Kiev.

_____ **2.** Ukraine and Belarus become republics of the Soviet Union.

_____ **3.** Greek Orthodox ministers teach Ukrainians and Belorussians about Christianity.

_____ **4.** The Soviet Union collapses, and Ukraine and Belarus declare independence.

_____ **5.** Viking traders establish the city of Kiev.

_____ **6.** The Greeks establish trading colonies along the coast of the Black Sea.

_____ **7.** All of modern Ukraine and Belarus come under Moscow's rule.

_____ **8.** The Slavs move into what is now Ukraine and Belarus.

Ukraine, Belarus, and the Caucasus

EASTERN CHAPTER 11

CHAPTER 18

SECTION 3

Reading the Section • As you read the section, use the space provided to identify the country described by each statement. Choose your answers from the list below. Answers will be used more than once.

Georgia   Armenia   Azerbaijan

_____ **1.** This country is a little smaller than Maryland and has fewer than 4 million people.

_____ **2.** This small country is located between the Caucasus Mountains and the Black Sea.

_____ **3.** The Azeri in this country speak a Turkic language.

_____ **4.** In 1988 an earthquake destroyed nearly one third of its industry.

_____ **5.** This country is mostly an agrarian society.

_____ **6.** Agriculture accounts for about 40 percent of its gross domestic product.

_____ **7.** Its major crops are citrus fruits and tea.

_____ **8.** Cotton, natural gas, and oil are this country's main products.

_____ **9.** Its sturgeon roe is made into some of the world's finest caviar.

_____**10.** Tourism on the Black Sea has helped this country's economy.

Post-Reading Quick Check • After you have finished reading the section, in the space provided, identify three problems that the Caucasus countries have faced in recent years.

1. Problem: _____

2. Problem: _____

3. Problem: _____

Central Asia

EASTERN CHAPTER 12

Reading the Section • As you read the section, match each name in the left column with its description in the right column. Write the letter of the correct description in the space provided. Answers may be used more than once.

_____ **1.** Fergana

_____ **2.** Hindu Kush

_____ **3.** Kyzyl Kum

_____ **4.** Caspian

_____ **5.** Pamirs

_____ **6.** Kara-Kum

_____ **7.** Syr Dar'ya

_____ **8.** Tian Shan

_____ **9.** Aral

_____ **10.** Amu Dar'ya

a. mountain

b. desert

c. river

d. valley

e. sea

Post-Reading Quick Check • After you have finished reading the section, in the space provided, describe the mineral resources of Central Asia.

_____

_____

_____

_____

_____

_____

_____

_____

People, Places, and Change

Central Asia

EASTERN CHAPTER 12

Reading the Section • As you read the section, answer each of
the following questions in the space provided.

1. Where did the best land route between China and the Mediterranean run? _____

2. Why did early merchants travel in caravans? _____

3. Why did Central Asia become isolated and poor in the 1500s? _____

4. Who conquered Central Asia in the 1200s? _____

5. When did the Soviets set up five republics in Central Asia? _____

6. What was the official language in Central Asia during the Soviet era? _____

7. When did the five Central Asian republics become independent? _____

8. To what alphabet are the five Central Asian countries switching? _____

9. What type of government rules the five Central Asian countries? _____

10. What does industry in Central Asia involve? _____

Post-Reading Quick Check • After you have finished reading
the section, in the space provided, list three benefits that Soviet
rule brought to Central Asia.

1. Benefit: _____

_____

2. Benefit: _____

_____

3. Benefit: _____

_____

CHAPTER 19 — Central Asia

EASTERN CHAPTER 12

Reading the Section • As you read the section, examine the descriptions of Central Asian countries below. For each description, place a check mark in the box next to the country being described.

1. This country has the largest population of the Central Asian countries.
   ❏ Kazakhstan   ❏ Kyrgystan   ❏ Turkmenistan   ❏ Uzbekistan   ❏ Tajikistan

2. Clan membersip is important in this Central Asian country.
   ❏ Kazakhstan   ❏ Kyrgystan   ❏ Turkmenistan   ❏ Uzbekistan   ❏ Tajikistan

3. The yurt is a symbol of this country's nomadic heritage.
   ❏ Kazakhstan   ❏ Kyrgystan   ❏ Turkmenistan   ❏ Uzbekistan   ❏ Tajikistan

4. Many men in this country wear black and white felt hats that show their clan status.
   ❏ Kazakhstan   ❏ Kyrgystan   ❏ Turkmenistan   ❏ Uzbekistan   ❏ Tajikistan

5. This country's language is related to Persian.
   ❏ Kazakhstan   ❏ Kyrgystan   ❏ Turkmenistan   ❏ Uzbekistan   ❏ Tajikistan

6. This country was the first to be conquered by Russia.
   ❏ Kazakhstan   ❏ Kyrgystan   ❏ Turkmenistan   ❏ Uzbekistan   ❏ Tajikistan

7. This country's people consider the great Persian literature to be part of their heritage.
   ❏ Kazakhstan   ❏ Kyrgystan   ❏ Turkmenistan   ❏ Uzbekistan   ❏ Tajikistan

8. The people in this country are required to study Uzbek to be eligible for citizenship.
   ❏ Kazakhstan   ❏ Kyrgystan   ❏ Turkmenistan   ❏ Uzbekistan   ❏ Tajikistan

9. Russian and Kazakh are its official languages.
   ❏ Kazakhstan   ❏ Kyrgystan   ❏ Turkmenistan   ❏ Uzbekistan   ❏ Tajikistan

10. The government of this country has ordered the schools to teach Islamic principles.
    ❏ Kazakhstan   ❏ Kyrgystan   ❏ Turkmenistan   ❏ Uzbekistan   ❏ Tajikistan

Post-Reading Quick Check • After you have finished reading the section, in the space provided, describe the civil war in Tajikistan in the mid-1990s.

_____

_____

_____

_____

_____

The Arabian Peninsula, Iraq, Iran, and Afghanistan

EASTERN CHAPTER 13

SECTION 1

Reading the Section • As you read the section, match each term or place in the left column with its description in the right column. Write the letter of the correct description in the space provided.

_____ **1.** Arabian Peninsula

_____ **2.** Mesopotamia

_____ **3.** Afghanistan

_____ **4.** Iran

_____ **5.** Yemen

_____ **6.** Hindu Kush

_____ **7.** Saudi Arabia

_____ **8.** wadis

_____ **9.** fossil water

_____ **10.** Persian Gulf

a. Country in which the Arabian Peninsula reaches its highest point

b. Gulf whose shores are the location of most of the region's oil fields

c. Dry streambeds

d. Term for water that is not being replaced by rainfall

e. Country bordered by the Elburz and the Kopet-Dag in the north and the Zagros in the southwest

f. Ancient name given to the area between the Tigris and Euphrates Rivers

g. Large rectangular area bordered by the Red Sea, Gulf of Aden, Arabian Sea, and Persian Gulf

h. Mountain range in Afghanistan

i. Country in which the Hindu Kush range can be found

j. Country that contains the largest sand desert in the world

Post-Reading Quick Check • After you have finished reading the section, in the space provided, describe the mineral resources of the region.

_____

_____

_____

_____

_____

_____

_____

Name _____ Class _____ Date _____

The Arabian Peninsula, Iraq, Iran, and Afghanistan

CHAPTER 20

SECTION 2

EASTERN CHAPTER 13

Reading the Section • As you read the section, complete the graphic organizer by providing information about Saudi Arabia in each of the categories shown.

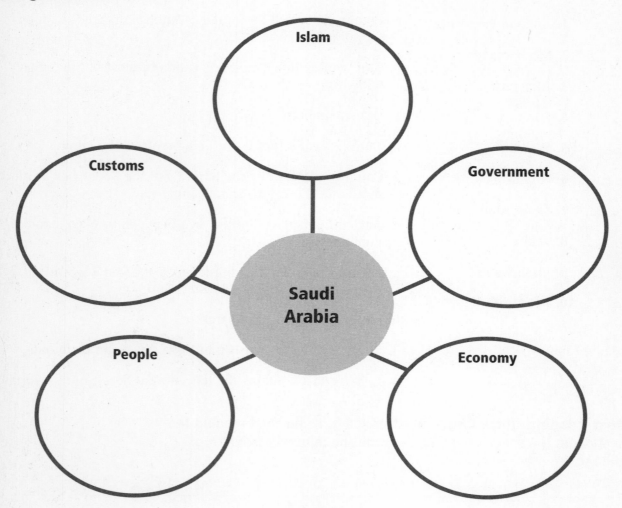

Islam

Customs

Government

Saudi Arabia

People

Economy

Post-Reading Quick Check • After you have finished reading the section, in the space provided, identify the six coastal countries that share the Arabian Peninsula with Saudi Arabia.

1. _____

2. _____

3. _____

4. _____

5. _____

6. _____

The Arabian Peninsula, Iraq, Iran, and Afghanistan

SECTION 3

EASTERN CHAPTER 13

Reading the Section • As you read the section, number the following events in the order in which they occurred.

_____ **1.** The Ba'ath Party takes power in Iraq.

_____ **2.** Saddam Hussein becomes president of Iraq and leader of the armed forces.

_____ **3.** A group of Iraqi officers overthrow the kingdom of Iraq set up by the British.

_____ **4.** Alexander the Great makes Mesopotamia part of his empire.

_____ **5.** The Persians conquer Mesopotamia.

_____ **6.** Mesopotamia becomes part of the Ottoman Empire.

_____ **7.** Iraq, led by Saddam Hussein, invades Iran.

_____ **8.** Mesopotamia is conquered by the Arabs.

_____ **9.** Iraq invades the small, oil-rich country of Kuwait.

_____ **10.** The British establish the kingdom of Iraq and place a pro-British ruler on the throne.

Post-Reading Quick Check • After you have finished reading the section, in the space provided, explain why Iraq's economy has suffered in recent years.

_____

_____

_____

_____

_____

_____

_____

_____

The Arabian Peninsula, Iraq, Iran, and Afghanistan

SECTION 4

EASTERN CHAPTER 13

Reading the Section • As you read the section, consider each of the statements below. In the space provided, write "I" if the statement refers to Iran or "A" if the statement refers to Afghanistan.

_____ **1.** The Khyber Pass is located along its border with Pakistan.

_____ **2.** The British and Russian Empires fought for control here in the 1800s.

_____ **3.** A revolution in 1979 made Islam this country's guiding force.

_____ **4.** The Persian Empire was established here in the 500s B.C.

_____ **5.** In 1979 the Soviet Union sent troops here to help the government win a civil war.

_____ **6.** An alliance of groups took over the government here after the Soviets left in 1989.

_____ **7.** A military officer took power here in 1921 under the title of shah.

_____ **8.** The Taliban have ruled this country strictly since the mid-1990s.

_____ **9.** This country is a theocracy, meaning that it is ruled by religious leaders.

_____ **10.** This country has one of the largest populations in Southwest Asia.

Post-Reading Quick Check • After you have finished reading the section, in the space provided, discuss the economies of Iran and Afghanistan.

1. Iran: _____

_____

_____

_____

2. Afghanistan: _____

_____

_____

_____

People, Places, and Change

The Eastern Mediterranean

EASTERN CHAPTER 14

Reading the Section • As you read the section, examine each of the pairs of statements below. Circle the letter of the statement in each pair that is true.

1. a. The Occupied Territories controlled by Israel include Turkey, Syria, and Lebanon.
  b. The Occupied Territories controlled by Israel include the West Bank, Gaza Strip, and Golan Heights.

2. a. The eastern Mediterranean region straddles three continents—Asia, Europe, and Africa.
  b. The eastern Mediterranean region straddles two continents—Asia and Europe.

3. a. The Dardanelles, the Bosporus, and the Sea of Marmara separate Europe from Asia.
  b. The Tigris and Euphrates Rivers and the Red Sea separate Europe from Asia.

4. a. The Dead Sea separates the two main ridges in the region.
  b. The Jordan River valley separates the two main ridges in the region.

5. a. The Dead Sea is so salty that swimmers cannot sink in it.
  b. The Dead Sea is so salty that swimmers immediately sink in it.

6. a. Central Syria and lands farther south have a desert climate.
  b. Central Syria and lands farther south have a Mediterranean climate.

7. a. A desert known as the Negev lies in southern Israel.
  b. A desert known as the Negev lies in northern Turkey.

8. a. The countries of the eastern Mediterranean do not have large reserves of oil.
  b. The countries of the eastern Mediterranean have large reserves of oil.

Post-Reading Quick Check • After you have finished reading the section, in the space provided, hypothesize how the Dead Sea got its name.

_____

_____

_____

_____

_____

_____

The Eastern Mediterranean

EASTERN CHAPTER 14

Reading the Section • As you read the section, circle the boldface word or phrase that *best* completes each statement below.

1. In the 330s B.C. **Alexander the Great / Kemal Atatürk** conquered Asia Minor.

2. The Ottoman Turks made **Constantinople / Ankara** the capital of their empire.

3. In World War I the Ottoman Empire fought on the **winning / losing** side.

4. To help modernize Turkey, the Arabic alphabet was replaced by the **Greek / Latin** alphabet.

5. Turkey's legislature is known as the **National Assembly / Turkish Parliament**.

6. Most of the people living in Turkey today are **Muslim / Christian**.

7. Turkish law now permits men to have **up to four wives / only one wife**.

8. **Making clothing / Mining minerals** is the most important industry in Turkey today.

9. **Kurds / Ethnic Turks** make up the largest minority group in Turkey today.

10. Shish kebab, a type of **furniture / food**, is a favorite among the Turkish people.

Post-Reading Quick Check • After you have finished reading the section, in the space provided, explain why some Turks were unhappy with the efforts to modernize Turkey.

_____

_____

_____

_____

_____

_____

_____

_____

_____

_____

The Eastern Mediterranean

SECTION 3

EASTERN CHAPTER 14

Reading the Section • As you read the section, number the following events in the order in which they occurred.

_____ **1.** A movement called Zionism begins among European Jews.

_____ **2.** The Arabs conquer Palestine.

_____ **3.** Tens of thousands of Jews move to the area of Palestine.

_____ **4.** Palestine comes under the control of the British.

_____ **5.** Israel and its Arab neighbors fight a series of wars.

_____ **6.** Ancient Jews establish the kingdom of Israel.

_____ **7.** The Romans conquer an area that they call Palestine.

_____ **8.** The Crusaders capture Jerusalem.

_____ **9.** The Jews in Palestine form the State of Israel.

_____ **10.** The Romans force most Jews to leave the region, resulting in the Diaspora.

Post-Reading Quick Check • After you have finished reading the section, in the space provided, identify and describe the Occupied Territories controlled by Israel.

1. Territory: _____ Description: _____

_____

_____

2. Territory: _____ Description: _____

_____

_____

3. Territory: _____ Description: _____

_____

_____

Name _____ Class _____ Date _____

The Eastern Mediterranean

EASTERN CHAPTER 14

Reading the Section • As you read the section, use the space provided to identify the country described by each statement. Choose your answers from the list below. Answers will be used more than once.

    Syria   Lebanon   Jordan

_____ **1.** This country's borders were drawn by Great Britain.

_____ **2.** Its capital is believed to be the oldest continuously inhabited city in the world.

_____ **3.** Until 1949 it was known as Transjordan.

_____ **4.** This country is small and mountainous and is located on the Mediterranean coast.

_____ **5.** For centuries, this country was a leading regional trade center.

_____ **6.** During the Ottoman period, many religious and ethnic minority groups settled here.

_____ **7.** From 1952 until 1999 this country was ruled by King Hussein.

_____ **8.** The civil war that occurred here from the 1970s until 1990 killed tens of thousands of people and damaged the capital.

_____ **9.** The refining of crude oil is a leading industry here now.

_____ **10.** This country is rich in limestone, basalt, and phosphates.

Post-Reading Quick Check • After you have finished reading the section, in the space provided, identify the capitals of Syria, Lebanon, and Jordan.

1. Syria: _____

2. Lebanon: _____

3. Jordan: _____

North Africa

EASTERN CHAPTER 15

SECTION 1

Reading the Section • As you read the section, complete the following outline by supplying the main idea and the missing subtopics and supporting details.

Physical Geography

Main Idea: _____

_____

Topic I: North Africa's physical features include a desert, mountains, and waterways.

Detail A: The Sahara, which covers much of North Africa, includes ergs and regs.

Detail B: _____

Detail C: The world's longest river, the Nile, flows northward through eastern Sahara.

Detail D: _____

Topic II: _____

Detail A: A desert climate covers most of the region, where plants and animals live.

Detail B: _____

Detail C: _____

Topic III: North Africa has important resources.

Detail A: Good soils and rain allow farmers to grow a variety of crops.

Detail B: _____

Detail C: Mineral resources include iron ore, copper, gold, and silver.

Post-Reading Quick Check • After you have finished reading the section, in the space provided, explain how the annual flooding of the Nile was important for farming.

_____

_____

_____

_____

North Africa

EASTERN CHAPTER 15

SECTION 2

Reading the Section • As you read the section, examine the riddles below. Solve each riddle by writing the correct word or place in the space provided.

_____ **1.** "We are the people who built great stone pyramids and buried our beloved pharaohs in them. Who are we?"

_____ **2.** "We are the pictures and symbols that the early Egyptians used to stand for ideas and words. We became the basis for Egypt's first writing system. What are we?"

_____ **3.** "Although Egypt gained some independence from us in 1922, we kept military bases there until 1956. Who are we?"

_____ **4.** "In 1979 Egypt became the first Arab country to sign a peace treaty with me. What country am I?"

_____ **5.** "We are nomadic herders who travel throughout the deserts of Egypt and southwest Asia. Who are we?"

_____ **6.** "I am the North African country that has the largest number of non-Muslims. In fact, about 6 percent of my people are Christians or members of other religions. What country am I?"

_____ **7.** "I am a food that is made from wheat and looks like tiny pellets of pasta. To get my full flavor, steam me over boiling water or soup. What food am I?"

_____ **8.** "I am the prophet of Islam. My followers celebrate my birthday with lights, parades, and special sweets. Who am I?"

Post-Reading Quick Check • After you have finished reading the section, in the space provided, explain how Islam and the Arabic language came to North Africa.

_____

_____

_____

_____

_____

_____

Name _____ Class _____ Date _____

North Africa

EASTERN CHAPTER 15

Reading the Section • As you read the section, complete the chart by providing information about Egypt today in each of the areas shown.

People and Cities
Rural Egypt:
Cities:
Economy
Industries:
Suez Canal:
Agriculture:
Challenges
Challenge:
Challenge:
Challenge:
Challenge:

Post-Reading Quick Check • After you have finished reading the section, in the space provided, identify three ways that the *fellahin* of Egypt support themselves.

1. _____

2. _____

3. _____

People, Places, and Change

North Africa

EASTERN CHAPTER 15

Reading the Section • All of the following statements refer to North African countries. As you read the section, use the space provided to indicate whether each statement below refers to Libya, Tunisia, Algeria, or Morocco.

_____ **1.** This country's city of Tangier overlooks the Strait of Gibraltar.

_____ **2.** This country is the only North African country that has little oil.

_____ **3.** Although this country is almost completely desert, it is the most urbanized country in the region.

_____ **4.** Farmers make up about half of the labor force here.

_____ **5.** About 80 percent of this country's trade is with EU countries.

_____ **6.** Algiers, with a population of 3.7 million, is this country's capital.

_____ **7.** This country has been ruled by the dictator Mu'ammar al-Gadhafi since 1969.

_____ **8.** Violence between this country's government and some Islamic groups has claimed thousands of lives.

_____ **9.** This country's largest cities are Tripoli, the capital, and Benghazi.

_____ **10.** The Casbah can be found in this country.

Post-Reading Quick Check • After you have finished reading the section, in the space provided, draw an arrow from each term in the left column to its definition in the right column.

1. Casbah   **a.** City in which almost no taxes are placed on goods sold there

2. souks   **b.** Marketplaces in Algiers

3. free port   **c.** Someone who rules a country with complete power

4. dictator   **d.** Old district of Algiers

West Africa

EASTERN CHAPTER 16

SECTION 1

Reading the Section • As you read the section, provide the following information about West Africa.

1. First climate zone: _____

2. Location: _____

3. Type of climate: _____

4. Second climate zone: _____

5. Location: _____

6. Type of climate: _____

7. Third climate zone _____

8. Location: _____

9. Type of climate: _____

10. Fourth climate zone: _____

11. Location: _____

12. Type of climate: _____

Post-Reading Quick Check • After you have finished reading the section, in the space provided, explain why the Niger River is so important to West Africa.

_____

_____

_____

_____

_____

_____

_____

Name _____ Class _____ Date _____

West Africa

EASTERN CHAPTER 16

Reading the Section • As you read the section, complete each sentence below by writing the appropriate word, name, or place in the space provided.

1. _____ is the study of the remains and ruins of past cultures.

2. The first known West African kingdom, called _____, became rich and powerful because of trade.

3. _____, the king of Mali in the 1300s, was wealthy and wise.

4. Songhay's city of _____ was a great cultural center, with a university, mosques, and more than 100 schools.

5. Europeans called the west coast of Africa the _____ because of the gold they bought there.

6. Europeans met the demand for labor in the American colonies by selling

   enslaved _____ to the colonists.

7. In the late 1800s only the country of _____ was able to remain independent of European control.

8. _____ was the last country to give up its West African colonies.

9. Today, most people of the Sahel practice the religion of _____.

10. Many homes in the Sahel and savanna zones are _____.

Post-Reading Quick Check • After you have finished reading the section, in the space provided, identify three problems that challenge West African countries today.

1. Problem: _____

   _____

2. Problem: _____

   _____

3. Problem: _____

   _____

People, Places, and Change

CHAPTER 23 West Africa

EASTERN CHAPTER 16

SECTION 3

Reading the Section • As you read the section, examine the descriptions of Sahel countries below. For each description, place a check mark in the box next to the country being described.

1. About 80 percent of this country's people fish or farm along the Niger River.
❑ Mauritania   ❑ Mali   ❑ Niger   ❑ Chad   ❑ Burkina Faso

2. This country's name means "land of the honest people."
❑ Mauritania   ❑ Mali   ❑ Niger   ❑ Chad   ❑ Burkina Faso

3. This country is located in the center of Africa.
❑ Mauritania   ❑ Mali   ❑ Niger   ❑ Chad   ❑ Burkina Faso

4. Many people who live here are Moors and speak Arabic.
❑ Mauritania   ❑ Mali   ❑ Niger   ❑ Chad   ❑ Burkina Faso

5. Only about 3 percent of this country's land is good for farming.
❑ Mauritania   ❑ Mali   ❑ Niger   ❑ Chad   ❑ Burkina Faso

6. This country's main export is cotton.
❑ Mauritania   ❑ Mali   ❑ Niger   ❑ Chad   ❑ Burkina Faso

7. This country has few mineral resources and thin soil.
❑ Mauritania   ❑ Mali   ❑ Niger   ❑ Chad   ❑ Burkina Faso

8. All of the farmland in this country lies along the Niger River.
❑ Mauritania   ❑ Mali   ❑ Niger   ❑ Chad   ❑ Burkina Faso

9. Lake Chad in this country is today only one third as large as it was in 1950.
❑ Mauritania   ❑ Mali   ❑ Niger   ❑ Chad   ❑ Burkina Faso

10. Its capital, Nouakchott, has grown to more than 700,000 people in just 40 years.
❑ Mauritania   ❑ Mali   ❑ Niger   ❑ Chad   ❑ Burkina Faso

Post-Reading Quick Check • After you have finished reading the section, in the space provided, discuss what the countries of the Sahel have in common.

_____

_____

_____

_____

People, Places, and Change

West Africa

CHAPTER 23

SECTION 4

EASTERN CHAPTER 16

Reading the Section • As you read the section, match each place in the left column with its description in the right column. Write the letter of the correct description in the space provided.

_____ **1.** Nigeria

_____ **2.** Senegal

_____ **3.** Gambia

_____ **4.** Guinea

_____ **5.** Guinea-Bissau

_____ **6.** Cape Verde

_____ **7.** Liberia

_____ **8.** Sierra Leone

_____ **9.** Ghana

_____ **10.** Côte d'Ivoire

a. Main resource is a huge supply of bauxite

b. Exports diamonds

c. West Africa's only island country

d. Largest country along West Africa's coast

e. Its name means "Ivory Coast"

f. Has very close ties with Senegal

g. Settled by freed American slaves

h. Its capital, Dakar, is an important seaport and manufacturing center

i. Has undeveloped mineral resources

j. Has one of the largest human-made lakes in the world—Lake Volta

Post-Reading Quick Check • After you have finished reading the section, in the space provided, list five facts about Togo and Benin.

1. Fact: _____

2. Fact: _____

3. Fact: _____

4. Fact: _____

5. Fact: _____

East Africa

EASTERN CHAPTER 17

Reading the Section • As you read the section, examine each of the pairs of statements below. Circle the letter of the statement in each pair that is true.

1. **a.** Mount Kilimanjaro is Africa's tallest mountain.
   b. Mount Everest is Africa's tallest mountain.

2. **a.** East Africa's most striking physical features are its great rifts.
   b. East Africa's most striking physical features are its great deserts.

3. **a.** The western rift extends from Lake Albert to Lake Malawi.
   b. The western rift extends from the Nile to the Mediterranean Sea.

4. **a.** The Nile, the world's shortest river, begins in East Africa.
   b. The Nile, the world's longest river, begins in East Africa.

5. **a.** The great Nile River is formed from the White Nile and the Blue Nile.
   b. The great Nile River is formed from the Red Nile and the Yellow Nile.

6. **a.** Lake Victoria is Africa's largest lake in area, but it is shallow.
   b. Lake Victoria is Africa's smallest lake in area, but it is deep.

7. **a.** Lake Victoria is too salty for most fish to live in it.
   b. Lake Nakuru is too salty for most fish to live in it.

8. **a.** East Africa has only one climate—a steppe climate.
   b. East Africa's climate is as varied as its landscape.

9. **a.** Most East Africans are farmers or herders.
   b. Most East Africans work in mining and manufacturing.

10. **a.** Among East Africa's many mineral resources are coal, copper, and diamonds.
    b. East Africa has few mineral resources.

Post-Reading Quick Check • After you have finished reading the section, in the space provided, define the term *rift* and explain how rifts are formed.

1. Definition: _____

_____

2. Explanation: _____

_____

_____

People, Places, and Change

East Africa

EASTERN CHAPTER 17

SECTION 2

Reading the Section • **As you read the section, circle the boldface word or phrase that *best* completes each statement below.**

1. About A.D. 400 Ethiopia's king conquered Sudan and adopted **Islam / Christianity**.

2. The **French / Portuguese** set up the first European forts and settlements on the East African coast in the early 1500s.

3. In the 1880s control over much of East Africa went to the **Americans / British**.

4. Within East Africa, just **Somalia / Kenya** was settled by large numbers of Europeans.

5. **Ethiopia / Burundi**, protected by its mountains, was never colonized.

6. Most East African countries were granted independence in the early **1920s / 1960s**.

7. East Africa has the world's **shortest / longest** history of human settlement.

8. Swahili is a **language / religion** widely **spoken / practiced** in East Africa.

9. The northern part of the region has conflicts between Muslims and **Jews / Christians**.

10. Rwanda and **Kenya / Burundi** have seen the worst ethnic conflict in the region.

Post-Reading Quick Check • **After you have finished reading the section, in the space provided, explain how Islam was introduced to East Africa.**

_____

_____

_____

_____

_____

_____

_____

_____

_____

_____

People, Places, and Change

East Africa

EASTERN CHAPTER 17

Reading the Section • As you read the section, answer each of the following questions in the space provided.

1. What European merchants traded on the Indian Ocean coast in the 1800s? _____

2. When did Kenya gain independence from Britain? _____

3. What two countries united to create Tanzania? _____

4. What famous archaeological site can be found in Tanzania? _____

5. Where can Mount Kilimanjaro be found? _____

6. What country ruled Rwanda and Burundi after World War I? _____

7. What two ethnic groups live in Rwanda and Burundi? _____

8. What country's economy collapsed in the 1970s? _____

9. What is the largest country in Africa? _____

10. Which desert makes up the northern half of Sudan? _____

Post-Reading Quick Check • After you have finished reading the section, in the space provided, compare the views of the British and the Kikuyu concerning land.

1. British view: _____

_____

_____

_____

2. Kikuyu view: _____

_____

_____

_____

East Africa

CHAPTER 24

SECTION 4

EASTERN CHAPTER 17

Reading the Section • As you read the section, match each description in the left column with the country it describes in the right column. Write the letter of the correct country in the space provided. Answers may be used more than once.

_____ **1.** Located on the Red Sea

_____ **2.** Most people who live here have the same culture, language, religion, and way of life

_____ **3.** Once was part of Ethiopia

_____ **4.** One of the world's poorest countries

_____ **5.** Land of deserts and dry savannas

_____ **6.** Small desert country located on the Bab al-Mandab

_____ **7.** Has been experiencing serious droughts over the past 30 years

_____ **8.** Livestock and animal hides are its main exports

_____ **9.** Its people include the Issa and the Afar

_____ **10.** Was leased to the French by the Ethiopian government

a. Ethiopia

b. Eritrea

c. Somalia

d. Djibouti

Post-Reading Quick Check • After you have finished reading the section, in the space provided, explain why the region that includes Ethiopia, Eritrea, Somalia, and Djibouti is called the Horn of Africa.

_____

_____

_____

_____

_____

_____

_____

Central Africa

CHAPTER 25

EASTERN CHAPTER 18

SECTION 1

Reading the Section • As you read the section, indicate whether each statement below is true or false by writing *T* or *F* in the space provided.

_____ **1.** The Pacific Ocean lies off the western coast of central Africa.

_____ **2.** The Congo Basin lies generally in the center of the central African region.

_____ **3.** Volcanic mountains can be found in northwestern Cameroon.

_____ **4.** The highest mountains in the region lie along the Eastern Rift Valley.

_____ **5.** Lake Tanganyika and Lake Malawi can be found in the Western Rift Valley.

_____ **6.** Central Africa lies along the equator and in the low latitudes.

_____ **7.** The Congo Basin and much of the Atlantic coast have a marine west coast climate.

_____ **8.** Central Africa's tropical rain forest has been completely cleared for farming.

_____ **9.** North and south of the Congo Basin are large areas with a tropical savanna climate.

_____ **10.** The high eastern mountains of central Africa enjoy a highland climate.

_____ **11.** Central African rivers are used to produce hydroelectric power.

_____ **12.** Central Africa has few important minerals.

Post-Reading Quick Check • After you have finished reading the section, in the space provided, identify and describe central Africa's two major river systems.

1. River: _____ Description: _____

_____

_____

2. River: _____ Description: _____

_____

_____

Central Africa

EASTERN CHAPTER 18

SECTION 2

Reading the Section • As you read the section, write the letter of the *best* choice in the space provided.

_____ **1.** Many people in the former French, Spanish, and Portuguese colonies are
  a. Roman Catholic. **c.** Muslim.
  b. Hindu. **d.** Buddhist.

_____ **2.** What is *Makossa*?
  a. a hair style
  b. a type of music
  c. the capital of Gabon
  d. a type of dessert

_____ **3.** About 2,000 years ago new peoples began to move into central Africa from
  a. eastern Africa.
  b. North America.
  c. western Africa.
  d. South America.

_____ **4.** The last European colony in central Africa was
  a. Angola. **c.** Belgian Congo.
  b. Gabon. **d.** Cameroon.

_____ **5.** English is the official language in
  a. São Tomé and Príncipe.
  b. Zambia and Malawi.
  c. Equatorial Guinea.
  d. Angola.

_____ **6.** What group of languages is spoken in most of central Africa today?
  a. Latin **c.** Bantu
  b. Portuguese **d.** Italian

_____ **7.** *Fufu* is a(n)
  a. sport.
  b. type of automobile.
  c. article of clothing.
  d. food.

_____ **8.** Where was the Kongo Kingdom located?
  a. at the mouth of the Zambezi River
  b. at the mouth of Lake Malawi
  c. at the mouth of Lake Tanganyika
  d. at the mouth of the Congo River

_____ **9.** When did the African colonies win their independence?
  a. after the Spanish-American War
  b. after World War I
  c. after World War II
  d. after the Korean War

_____ **10.** When did Europeans first arrive in central Africa?
  a. 600s **c.** 1400s
  b. 1200s **d.** 1900s

Post-Reading Quick Check • After you have finished reading the section, in the space provided, identify three challenges facing the people of central Africa today.

1. Challenge: _____

_____

2. Challenge: _____

_____

3. Challenge: _____

_____

Central Africa

CHAPTER 25

EASTERN CHAPTER **18**

Reading the Section • As you read the section, number the following events in the order in which they occurred.

_____ **1.** Mobutu Sese Seko changes the country's name to Zaire.

_____ **2.** The Congo Free State wins independence from Belgium.

_____ **3.** Portuguese sailors make contact with the Kongo Kingdom.

_____ **4.** Mobutu Sese Seko, a dictator, comes to power.

_____ **5.** King Leopold II of Belgium takes control of the Congo Basin.

_____ **6.** A new government takes over Zaire and renames it the Democratic Republic of the Congo.

_____ **7.** The Belgian government takes control of the Congo Free State.

_____ **8.** Many Belgians flee the Congo Free State after it wins independence.

Post-Reading Quick Check • After you have finished reading the section, in the space provided, supply the following information about the Democratic Republic of the Congo.

1. Population size: _____

2. Number of ethnic groups: _____

3. Official language: _____

4. Capital: _____

5. Resources: _____

6. Challenges: _____

_____

_____

_____

_____

_____

Central Africa

EASTERN CHAPTER 18

SECTION 4

Reading the Section • As you read the section, examine each of
the pairs of statements below. Circle the letter of the statement in
each pair that is true.

1. **a.** Equatorial Guinea gained independence from the United States in 1968.
   b. Equatorial Guinea gained independence from Spain in 1968.

2. **a.** Cameroon is by far the most populous country in northern central Africa.
   b. Gabon is by far the most populous country in northern central Africa.

3. **a.** The capital of Cameroon is Yaoundé.
   b. The capital of Cameroon is Brazzaville.

4. **a.** The Central African Republic has the strongest economy in the region.
   b. Gabon has the strongest economy in the region.

5. **a.** Many of the region's goods are shipped down the Congo River to Brazzaville.
   b. Many of the region's goods are shipped down the Zambezi River to Brazzaville.

6. **a.** Zambia and Malawi won independence from Great Britain in 1964.
   b. Zambia and Malawi won independence from Portugal in 1964.

7. **a.** The populations of the three countries in southern central Africa vary widely.
   b. The populations of the three countries in southern central Africa are nearly the same.

8. **a.** Angola's capital, Luanda, is the southern region's smallest city.
   b. Angola's capital, Luanda, is the southern region's largest city.

9. **a.** Much of Zambia's income comes from copper.
   b. Much of Zambia's income comes from oil.

10. **a.** Most of the people who live in Malawi are farmers.
    b. Most of the people who live in Malawi are nomads.

Post-Reading Quick Check • After you have finished reading the
section, in the space provided, discuss what has happened in Angola
since the time of its independence.

_____

_____

_____

_____

_____

CHAPTER 26

Southern Africa

EASTERN CHAPTER 19

Reading the Section • As you read the section, complete the graphic organizer by providing information about southern Africa in each of the categories shown.

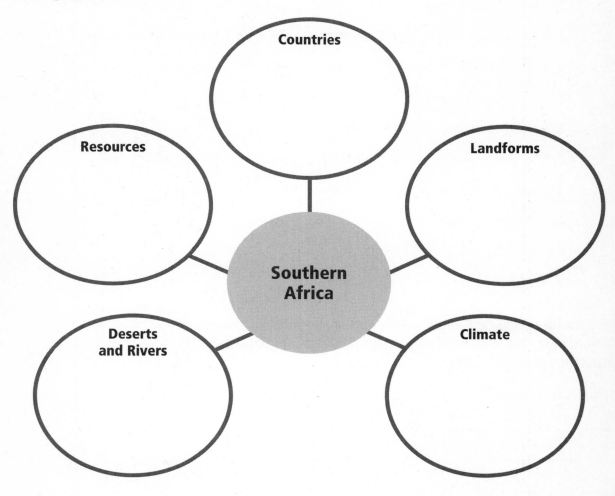

Countries

Resources

Landforms

Southern Africa

Deserts and Rivers

Climate

Post-Reading Quick Check • After you have finished reading the section, in the space provided, identify the four countries in southern Africa that are landlocked.

1. _____ 3. _____

2. _____ 4. _____

People, Places, and Change

Name _____ Class _____ Date _____

Southern Africa

CHAPTER 26

EASTERN CHAPTER 19

Reading the Section • As you read the section, number the
following events in the order in which they occurred.

_____ **1.** The Portuguese set up forts in Mozambique.

_____ **2.** The Shona build an empire in what is now Zimbabwe and Mozambique.

_____ **3.** Diamonds are discovered in the northern Cape Colony.

_____ **4.** Bantu-speaking people spread from central Africa into southern Africa.

_____ **5.** Great Britain bans slavery in its empire.

_____ **6.** The Dutch set up a trade station at a natural harbor near Cape Town.

_____ **7.** Zimbabwe becomes the British colony of Southern Rhodesia.

_____ **8.** Tensions between Boers and the British lead to war.

_____ **9.** Botswana comes under British control.

_____ **10.** People from Asia settle on Madagascar.

Post-Reading Quick Check • After you have finished reading the
section, in the space provided, explain how the language called
Afrikaans came into being.

_____

_____

_____

_____

_____

_____

_____

_____

_____

People, Places, and Change

Southern Africa

CHAPTER 26

EASTERN CHAPTER 19

SECTION 3

Reading the Section • As you read the section, complete the following outline by supplying the main idea and the missing subtopics and supporting details.

South Africa Today

Main Idea: _____

_____

Topic I: South Africa had a long history of racist laws.

Detail A: Dominated by Afrikaners, South Africa's government became increasingly racist in the early 1900s.

Detail B: _____

Detail C: Whites owned most of the good farmland, mines, and other natural resources.

Detail D: _____

Topic II: _____

Detail A: Other countries in southern Africa gained their independence.

Detail B: _____

Detail C: Today, all races in South Africa have equal rights.

Topic III: South Africa has important resources, but still has problems.

Detail A: Resources include coal, hydroelectric power, diamonds, and other minerals.

Detail B: _____

Detail C: _____

Post-Reading Quick Check • After you have finished reading the section, in the space provided, identify three ways that people around the world protested apartheid.

1. Protest: _____

2. Protest: _____

3. Protest: _____

Southern Africa

CHAPTER 26

EASTERN CHAPTER 19

SECTION 4

Reading the Section • As you read the section, use the space provided to identify the country described by each statement. Choose your answers from the list below. Answers will be used more than once.

Namibia　　　　Botswana　　　　Zimbabwe

Mozambique　　　　Madagascar

_____ **1.** Its capital was called Salisbury when the country was known as Southern Rhodesia.

_____ **2.** The white people who live here are mainly of German descent.

_____ **3.** This country was once a French colony.

_____ **4.** Traditional crafts in this country include ostrich-eggshell beadwork.

_____ **5.** About 95 percent of the people here belong to the Tswana ethnic group.

_____ **6.** It is one of the world's poorest countries.

_____ **7.** This country exports tobacco, corn, sugar, and beef.

_____ **8.** Most of the people in this country are Christians, and English is the official language.

_____ **9.** Its economy was badly damaged by civil war.

_____ **10.** There is little industry in this country.

Post-Reading Quick Check • After you have finished reading the section, in the space provided, draw an arrow from each country in the left column to its capital in the right column.

1. Namibia　　　　**a.** Harare

2. Botswana　　　　**b.** Windhoek

3. Zimbabwe　　　　**c.** Maputo

4. Mozambique　　　　**d.** Gaborone

China, Mongolia, and Taiwan

CHAPTER 27

EASTERN CHAPTER 20

Reading the Section • Each of the following sentences contains an underlined word or name that makes the sentence incorrect. As you read the section, use the space provided to write the word or name that makes the sentence correct.

_____ 1. <u>Taiwan</u> has some of the world's highest mountains, driest deserts, and longest rivers.

_____ 2. <u>Mongolia</u> is a tropical island just off mainland China's coast.

_____ 3. The towering <u>Kunlan Shan</u>, the world's tallest mountain range, run along China's southwestern border.

_____ 4. The huge <u>Plain</u> of Tibet lies between the Himalayas and the Kunlan Mountains.

_____ 5. The <u>Sahara</u> Desert takes up much of the central and southeastern sections of the Mongolian Plateau.

_____ 6. The <u>South</u> China Plain is the largest plain in China.

_____ 7. The Huang, which means "<u>green</u> river," gets its name from the loess it carries.

_____ 8. The <u>Xi</u> is China's—and Asia's—longest river.

_____ 9. China has greater <u>gold</u> reserves than any other country in the world.

_____ 10. Arable land is the most important resource in <u>Mongolia</u>.

Post-Reading Quick Check • After you have finished reading the section, in the space provided, describe China's climate.

_____

_____

_____

_____

_____

_____

China, Mongolia, and Taiwan

CHAPTER 27

SECTION 2

EASTERN CHAPTER 20

Reading the Section • As you read the section, complete each sentence below by writing the appropriate word, name, or place in the space provided.

1. Beginning about 2000 B.C. northern Chinese living in the _____ valley formed kingdoms.

2. The Great Wall of China was begun under the _____ dynasty.

3. During the Han dynasty the Chinese invented the _____.

4. In the 1200s China was conquered by the _____.

5. In 1912 a revolutionary group led by _____ forced the last Chinese emperor to give up power.

6. Chiang Kai-shek and the Nationalists went to _____, where they created a government called the Republic of China.

7. Mao Zedong began a movement called the _____ Revolution that attempted to make all Chinese live a peasant way of life.

8. _____ Chinese is the official language of China.

9. The teachings of the philosopher _____ stressed family values.

10. An Indian prince named Siddhartha Gautama founded the religion known as

_____.

Post-Reading Quick Check • After you have finished reading the section, in the space provided, explain how Maoism changed life in China.

_____

_____

_____

_____

_____

_____

_____

China, Mongolia, and Taiwan

EASTERN CHAPTER 20

SECTION 3

Reading the Section • As you read the section, answer each of the following questions in the space provided.

1. What country has the largest population in the world? _____

2. What percentage of Chinese live in the western half of the country? _____

3. How many cities in China have populations greater than 1 million? _____

4. What is the capital of China? _____

5. What is the largest city in China? _____

6. What city, southeast of Guangzhou, is a former British colony? _____

7. What type of economy did the Chinese Communists set up in 1949? _____

8. What percentage of China's land is good for farming? _____

9. What are northern China's main crops? _____

10. What advantage does China's most-favored-nation status bring it? _____

Post-Reading Quick Check • After you have finished reading the section, in the space provided, identify two reasons why China is a world leader in agricultural production.

1. Reason: _____

_____

_____

_____

2. Reason: _____

_____

_____

_____

China, Mongolia, and Taiwan

CHAPTER 27

EASTERN CHAPTER 20

SECTION 4

Reading the Section • As you read the section, consider each of the statements listed below. In the space provided, write "M" if the statement refers to Mongolia or "T" if the statement refers to Taiwan.

_____ **1.** This country's vast empire reached its height in the 1200s.

_____ **2.** This island country once was known in the West as Formosa.

_____ **3.** The search for spices brought Europeans to this country.

_____ **4.** In 1911 this country, with Russian support, declared its independence from China.

_____ **5.** In 1949 Chiang Kai-shek and the Nationalist Chinese government fled to this country.

_____ **6.** Many people who live here still follow a nomadic way of life.

_____ **7.** This country has a modern, industrial economy and a population of nearly 22 million.

_____ **8.** The capital of this country is Taipei, located on the coastal plain.

_____ **9.** The nomads who live here make their homes in *gers*.

_____ **10.** This country's capital city is Ulaanbaatar.

Post-Reading Quick Check • After you have finished reading the section, in the space provided, describe Taiwan's industrial development.

_____

_____

_____

_____

_____

_____

_____

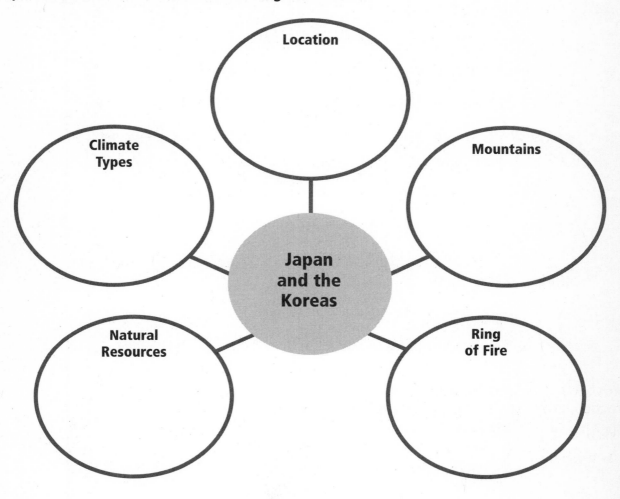

CHAPTER 28 — Japan and the Koreas

EASTERN CHAPTER 21

Reading the Section • As you read the section, complete the graphic organizer by supplying information about the physical geography of Japan and the Koreas in each of the categories shown.

Location

Climate Types

Mountains

Japan and the Koreas

Natural Resources

Ring of Fire

Post-Reading Quick Check • After you have finished reading the section, in the space provided, identify and describe the two types of current found in the region.

1. Current: _____ Description: _____

_____

2. Current: _____ Description: _____

_____

Japan and the Koreas

SECTION 2

EASTERN CHAPTER 21

Reading the Section • As you read the section, number the following events in the order in which they occurred.

_____ **1.** Japan takes control of Korea.

_____ **2.** Europeans are forced to leave Japan.

_____ **3.** Japan begins to expand its empire as a way to gain needed resources.

_____ **4.** Rice farming is introduced to Japan from China and Korea.

_____ **5.** Japan becomes an ally of Germany and Italy in World War II.

_____ **6.** Japan establishes a democratic government.

_____ **7.** The United States enters World War II after Japan attacks Pearl Harbor.

_____ **8.** Japan begins to develop a political system of its own.

_____ **9.** Japan begins its modernization period.

_____ **10.** Japan's first inhabitants come to Japan from central Asia.

_____ **11.** U.S. commodore Matthew Perry arrives with his warships in Tokyo Bay.

_____ **12.** Portuguese traders arrive in Japan.

Post-Reading Quick Check • After you have finished reading the section, in the space provided, identify and describe four religions that have been practiced in Japan.

1. Religion: _____ Description: _____

_____

2. Religion: _____ Description: _____

_____

3. Religion: _____ Description: _____

_____

4. Religion: _____ Description: _____

_____

People, Places, and Change

CHAPTER 28

Japan and the Koreas

EASTERN CHAPTER 21

Reading the Section • As you read the section, circle the boldface word or phrase that *best* completes each statement below.

1. Japan is one of the world's **least / most densely** populated countries.

2. **Only about 11 percent / Nearly 100 percent** of Japan's land is fit for growing crops.

3. Almost **30 million / 300 million** people live within 20 miles of the Imperial Palace in Tokyo.

4. **Osaka / Tokyo** is Japan's capital and the center of government.

5. **Over 99 percent / Less than 50 percent** of the Japanese population is ethnically Japanese.

6. Most Japanese sleep on lightweight cotton mattresses called **futons / kimonos**.

7. Japan produces about a third of its energy through **coal / nuclear power**.

8. Japan's **steel industry / fishing industry** is the largest in the world.

9. The average farm in Japan is **smaller / larger** than the average U.S. farm.

10. Japanese farmers make the most of their land by using **terracing / protectionism**.

Post-Reading Quick Check • After you have finished reading the section, in the space provided, identify the three major cities in Japan's Kansai region and provide one fact about each.

1. City: _____ Fact: _____

_____

2. City: _____ Fact: _____

_____

3. City: _____ Fact: _____

_____

Japan and the Koreas

EASTERN CHAPTER 21

SECTION 4

Reading the Section • All of the following statements refer to the history of Korea. As you read the section, use the space provided to indicate whether each statement below refers to ancient Korea, early modern Korea, or Korea since World War II.

_____ **1.** China closes Korea to most other outsiders, and Korea becomes known as the Hermit Kingdom.

_____ **2.** Korea adopts rice farming, which was introduced from China.

_____ **3.** Catholicism is introduced into Korea.

_____ **4.** North Korea becomes known as the Democratic People's Republic of Korea.

_____ **5.** Japan defeats China in the Sino-Japanese War.

_____ **6.** The Chinese introduce to Korea their writing system, their system of examinations for government jobs, and their religious traditions.

_____ **7.** North Korea invades South Korea, sparking the Korean War.

_____ **8.** During Korea's golden age, the country becomes well known in Asia for its architecture, painting, ceramics, and fine jewelry.

_____ **9.** North Korea and South Korea are divided by the demilitarized zone.

_____ **10.** The Christian community in Korea is small and sometimes faces persecution.

Post-Reading Quick Check • After you have finished reading the section, in the space provided, explain why Korea was divided into North Korea and South Korea.

_____

_____

_____

_____

_____

_____

Japan and the Koreas

EASTERN CHAPTER 21

Reading the Section • As you read the section, examine the riddles below. Solve each riddle by writing the correct place or term in the space provided.

_____ 1. "After the war, my population exploded because refugees came to me looking for housing and jobs. What city am I?"

_____ 2. "I am the type of government now found in South Korea. What type of government am I?"

_____ 3. "I am the most common religion found in South Korea today. What religion am I?"

_____ 4. "We link Korean businesses through family and personal ties, and we are huge. What are we?"

_____ 5. "I am Korea's most important crop—Korean farmers grow me on about half their land. What crop am I?"

_____ 6. "Because I am so yummy-tasting, Korea has made me its national dish. Guess what I am and I'll give you a taste."

_____ 7. "I am the party that controls the government of North Korea. What party am I?"

_____ 8. "We are made up of groups of North Korean farmers that work the land together. What are we?"

Post-Reading Quick Check • After you have finished reading the section, in the space provided, explain why South Korea is more prosperous than North Korea.

_____

_____

_____

_____

_____

_____

Southeast Asia

EASTERN CHAPTER 22

Reading the Section • As you read the section, complete the
following outline by supplying the main idea and the missing
subtopics and supporting details.

Physical Geography

Main Idea: _____

_____

Topic I: Physical features include peninsulas, archipelagos, mountains, and rivers.

Detail A: The Indochina and Malay Peninsulas lie on the Asian mainland.

Detail B: _____

Detail C: Both the mainland and the islands contain high mountains.

Detail D: _____

Topic II: _____

Detail A: The region tends to be warm year-round, but the higher elevations are cooler.

Detail B: _____

Detail C: The region is home to a wide variety of animals, many of which are endangered.

Topic III: Southeast Asia has a wide variety of resources.

Detail A: The rain forests provide many kinds of valuable wood.

Detail B: _____

Detail C: _____

Post-Reading Quick Check • After you have finished reading the
section, in the space provided, explain why the plants and animals of
the tropical rain forests are endangered.

_____

_____

_____

_____

People, Places, and Change

Southeast Asia

EASTERN CHAPTER 22

SECTION 2

Reading the Section • As you read the section, number the following events in the order in which they occurred.

_____ **1.** Japan invades and occupies much of the region during World War II.

_____ **2.** Wars in Southeast Asia end in communist victories.

_____ **3.** The United States grants independence to the Philippines.

_____ **4.** Europeans begin to establish colonies in Southeast Asia.

_____ **5.** European rule has ended in most of the region.

_____ **6.** The Portuguese, British, Dutch, French, and Spanish control most of the region.

_____ **7.** Indonesia allows the East Timorese to vote for independence.

_____ **8.** The Khmer Empire, based in what is now Cambodia, controls a large area.

_____ **9.** The French leave French Indochina, and their former colonies become independent.

_____ **10.** The United States wins control of the Philippines from Spain.

Post-Reading Quick Check • After you have finished reading the section, in the space provided, list the following information about Southeast Asia.

1. Religion of the region's Indian communities: _____

2. Main religion of the mainland countries: _____

3. Major religion in Malaysia, Brunei, and Indonesia: _____

4. World's largest Islamic country: _____

5. Major religion in the Philippines: _____

6. Most important food in the region: _____

7. Another important food: _____

Southeast Asia

CHAPTER 29

EASTERN CHAPTER 22

SECTION 3

Reading the Section • As you read the section, examine the descriptions of mainland Southeast Asian countries below. For each description, place a check mark in the box next to the country being described.

1. This mountainous, landlocked country has few good roads and no railroads.
   ❏ Vietnam   ❏ Laos   ❏ Cambodia   ❏ Thailand   ❏ Myanmar

2. This former British colony gained independence in 1948.
   ❏ Vietnam   ❏ Laos   ❏ Cambodia   ❏ Thailand   ❏ Myanmar

3. Among the mainland countries, this country has the strongest economy.
   ❏ Vietnam   ❏ Laos   ❏ Cambodia   ❏ Thailand   ❏ Myanmar

4. The Hong and Mekong River deltas are major farming areas in this country.
   ❏ Vietnam   ❏ Laos   ❏ Cambodia   ❏ Thailand   ❏ Myanmar

5. Until 1989 this country was known as Burma.
   ❏ Vietnam   ❏ Laos   ❏ Cambodia   ❏ Thailand   ❏ Myanmar

6. Agriculture is the most important part of this country's economy.
   ❏ Vietnam   ❏ Laos   ❏ Cambodia   ❏ Thailand   ❏ Myanmar

7. This country's economy has been slowly recovering since the end of the war in 1975.
   ❏ Vietnam   ❏ Laos   ❏ Cambodia   ❏ Thailand   ❏ Myanmar

8. Factories in this country produce electronics and computers.
   ❏ Vietnam   ❏ Laos   ❏ Cambodia   ❏ Thailand   ❏ Myanmar

9. Its capital and largest city, Phnom Penh, is located along the Mekong River.
   ❏ Vietnam   ❏ Laos   ❏ Cambodia   ❏ Thailand   ❏ Myanmar

10. The capital of this communist country is located in Vientiane.
    ❏ Vietnam   ❏ Laos   ❏ Cambodia   ❏ Thailand   ❏ Myanmar

Post-Reading Quick Check • After you have finished reading the section, in the space provided, explain why many Southeast Asians are moving to the cities.

_____

_____

_____

Southeast Asia

CHAPTER 29

SECTION 4

EASTERN CHAPTER 22

Reading the Section • As you read the section, match each description in the left column with the correct country in the right column. Write the letter of the country in the space provided. Answers may be used more than once.

_____ **1.** Its government is trying to attract more high-technology companies here

_____ **2.** A few very rich people control most of its land and industries

_____ **3.** Large deposits of oil have made it a very rich country

_____ **4.** Once called the Spice Islands because of its cinnamon, pepper, and other valuable spices

_____ **5.** The most economically developed country in Southeast Asia

_____ **6.** Most farmers here are poor and own no land

_____ **7.** It is ruled by a sultan

_____ **8.** Many foreign companies have opened banks, offices, and high-technology industries here

_____ **9.** The world's leading producer of palm oil

_____ **10.** Large areas of its rain forest often are burned for farming

a. Indonesia

b. Philippines

c. Singapore

d. Malaysia

e. Brunei

Post-Reading Quick Check • After you have finished reading the section, in the space provided, explain why crime rates are low in Singapore.

_____

_____

_____

_____

_____

India

CHAPTER 30

EASTERN CHAPTER 23

Reading the Section • As you read the section, answer each of the following questions in the space provided.

1. What is a subcontinent? _____

2. What mountains run along India's northern border? _____

3. How were the Himalayas created? _____

4. What two low mountain ranges define the edges of the Deccan plateau? _____

_____

5. What is India's most important river? _____

6. What do Hindus call the Ganges? _____

7. What desert is near India's border with Pakistan? _____

8. What type of climate can be found on the Gangetic Plain? _____

9. What type of work do most people in India do? _____

10. What mineral resources does India have? _____

Post-Reading Quick Check • After you have finished reading the section, in the space provided, identify India's three main landform regions.

1. Landform region: _____

2. Landform region: _____

3. Landform region: _____

CHAPTER 30 India

EASTERN CHAPTER 23

SECTION 2

Reading the Section • As you read the section, complete each sentence below by writing the appropriate word, name, or place in the space provided.

1. The first civilization on the Indian subcontinent was centered around the

   _____ River valley.

2. The _____ came into northern India by about 1500 B.C.

3. In the early 1200s a Muslim kingdom, called the Delhi _____, was established at Delhi.

4. The founder of the Mughal Empire was _____, whose name meant "the Tiger."

5. The reign of _____ and his successors was a golden age for India.

6. _____ had the Taj Mahal built as a tomb for his beloved wife.

7. During the 1700s and 1800s the _____ gradually took control of India.

8. The army of the British East India Company was made up mostly of

   _____, Indian troops commanded by British officers.

9. Mohandas K. _____ became the most important leader of the Indian independence movement.

10. In 1947 the British divided their Indian colony into two independent countries—

    _____ and _____.

Post-Reading Quick Check • After you have finished reading the section, in the space provided, describe the strategy used by the leader of the Indian independence movement.

_____

_____

_____

_____

CHAPTER 30

India

EASTERN CHAPTER 23

Reading the Section • As you read the section, complete the graphic organizer by supplying information about India in each of the categories shown.

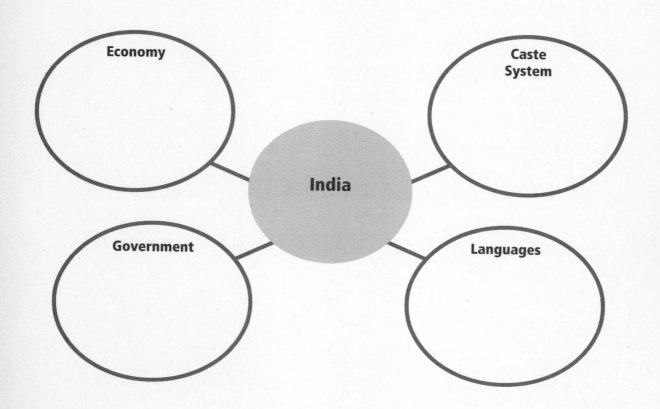

Post-Reading Quick Check • After you have finished reading the section, in the space provided, identify the four religions that originated in India.

1. _____

2. _____

3. _____

4. _____

Name _____ Class _____ Date _____

The Indian Perimeter

EASTERN CHAPTER 24

Reading the Section • As you read the section, examine each of the pairs of statements below. Circle the letter of the statement in each pair that is true.

1. **a.** The Indian Perimeter consists of all the large cities found in India.
   b. The Indian Perimeter consists of the countries that surround India.

2. **a.** North of Bangladesh is Bhutan, a tiny country high in the Himalayas.
   b. South of Bangladesh is Bhutan, a tiny country high in the Himalayas.

3. **a.** The Himalayas occupy some 10 percent of Nepal's land area.
   b. The Himalayas occupy some 80 percent of Nepal's land area.

4. **a.** Mount Everest, Earth's highest mountain, is located on Nepal's border with China.
   b. Hindu Kush, Earth's highest mountain, is located on Nepal's border with China.

5. **a.** Invaders and traders traveled through the Kyber Pass to the Indian interior.
   b. Invaders and traders traveled through the Tarai to the Indian interior.

6. **a.** Pakistan has one of the wettest climates in the world.
   b. Bangladesh has one of the wettest climates in the world.

7. **a.** Storm surges are huge waves of water whipped up by fierce winds.
   b. Storm surges are powerful winds that accompany cyclones.

8. **a.** Pakistan has large natural gas reserves.
   b. Pakistan has large deposits of diamonds and other gemstones.

9. **a.** Sri Lanka is a large peninsula located just off the southeastern tip of India
   b. Sri Lanka is a large island located just off the southeastern tip of India.

10. **a.** All of the islands in the Maldives are inhabited.
    b. Only about 200 of the 1,200 islands in the Maldives are inhabited.

Post-Reading Quick Check • After you have finished reading the section, in the space provided, explain how monsoons affect the weather of the northern Indian Perimeter.

_____

_____

_____

_____

_____

People, Places, and Change

The Indian Perimeter

EASTERN CHAPTER 24

SECTION 2

Reading the Section • As you read the section, circle the boldface word or phrase that *best* completes each statement below.

1. An ancient civilization developed in the **Ganges / Indus River** valley about 2500 B.C.

2. In 1947 Great Britain divided India into two countries, based on **race / religion**.

3. In 1971 East Pakistan changed its name to **Bangladesh / Nepal**.

4. Pakistan today is a largely **Christian / Islamic** country.

5. Pakistan shares with India and Bangladesh one of the world's **highest / lowest** birthrates.

6. Bangladesh is part of a region known as **Bengal / Borneo**.

7. **Hindus / Muslims** make up the majority of the population in Bangladesh.

8. Bangladesh's capital and largest city is **Islamabad / Dhaka**.

9. Marriages in Bangladesh are arranged by the **parents / couple**.

10. **English / Bengali** is the official language in Bangladesh and the language of the schools.

Post-Reading Quick Check • After you have finished reading the section, in the space provided, explain what contributes to the high rate of disease in Bangladesh.

_____

_____

_____

_____

_____

_____

_____

_____

The Indian Perimeter

EASTERN CHAPTER 24

SECTION 3

Reading the Section • As you read the section, match each description in the left column with the correct country in the right column. Write the letter of the country in the space provided. Answers may be used more than once.

_____ **1.** In the 1600s a Tibetan monk helped organize this country as a unified state

_____ **2.** There are no colleges or universities here

_____ **3.** The Buddha is thought to have been born here

_____ **4.** This country once was known as Ceylon

_____ **5.** Until the mid-1970s its government followed a policy of near total isolation

_____ **6.** This country's best farmland is found on the Tarai

_____ **7.** This archipelago consists of a series of 19 atolls

_____ **8.** Among the peoples of Tibetan origin here, the best known are the Sherpas

_____ **9.** This country leads the world in graphite exports

_____ **10.** Timber and hydroelectricity are among its most important resources

a. Nepal

b. Bhutan

c. Sri Lanka

d. Maldives

Post-Reading Quick Check • After you have finished reading the section, in the space provided, identify the capital of each of the following countries.

1. Nepal: _____

2. Bhutan: _____

3. Sri Lanka: _____

4. Maldives: _____

Australia and New Zealand

CHAPTER 32

EASTERN CHAPTER 25

SECTION 1

Reading the Section • As you read the section, examine the riddles below. Solve each riddle by writing the correct word or words in the space provided.

_____ **1.** "I am the world's smallest, flattest, and lowest continent. In addition, I am the only continent that is also a country. What am I?"

_____ **2.** "As part of the Eastern Highlands, I divide Australia's rivers into those that flow eastward and those that flow westward. What am I?"

_____ **3.** "The fact that I stretch more than 1,200 miles makes me the world's largest coral reef. Many kinds of marine life live in me. What am I?"

_____ **4.** "I am Australia's most common tree. I am large and leafy. What kind of tree am I?"

_____ **5.** "We were the first people to live in Australia. We moved there from Southeast Asia at least 40,000 years ago.Who are we?"

_____ **6.** "We began settling colonies in Australia near the end of the 1700s. Many of the first settlers came from our prisons. Who are we?"

_____ **7.** "I am the national capital of Australia. You can find me in the southeast part of the country. What city am I?"

_____ **8.** "Even though I am a British sport, Australians love to play me. I'm a lot like football and soccer. What sport am I?"

_____ **9.** "I am Australia's largest city. If you come for a visit, make sure you go see my beautiful Opera House. What city am I?"

_____ **10.** "I am the term used for Australia's lightly populated wilderness areas. Only about 15 percent of Australians live in these areas. What term am I?"

Post-Reading Quick Check • After you have finished reading the section, in the space provided, define the term *marsupial* and identify two of Australia's most well-known marsupials.

1. Definition: _____

_____

2. Marsupial: _____

3. Marsupial: _____

Australia and New Zealand

CHAPTER 32

EASTERN CHAPTER 25

SECTION 2

Reading the Section • As you read the section, write the correct word or place next to its description. Choose your answers from the list below. Some items will not be used.

Great Britain   Christchurch   Auckland
North Island   Spain   Southern Alps
wool   Wellington   *whare whakairo*
United States   Maori   kiwi

_____ 1. New Zealand's highest mountains, located on South Island

_____ 2. Flightless bird endemic to New Zealand

_____ 3. Descendants of New Zealand's first settlers

_____ 4. Country that granted New Zealand independence in 1907

_____ 5. Capital of New Zealand, located at the southern tip of North Island

_____ 6. Maori meeting houses

_____ 7. New Zealand's largest city and seaport

_____ 8. South Island's largest city

_____ 9. One of New Zealand's major products

_____ 10. One of New Zealand's main trading partners

Post-Reading Quick Check • After you have finished reading the section, in the space provided, describe the culture of New Zealand.

_____

_____

_____

_____

_____

_____

People, Places, and Change

The Pacific Islands and Antarctica

SECTION 1

EASTERN CHAPTER 26

Reading the Section • As you read the section, complete each sentence below by writing the appropriate word or phrase in the space provided.

1. The Pacific Ocean covers more than _____ of the surface of Earth.

2. The Pacific Islands are divided into three regions: Melanesia, _____, and Polynesia.

3. There are two kinds of high islands: _____ and continental.

4. Continental high islands were formed from continental _____.

5. The only island larger than _____ is Greenland.

6. The western half of New Guinea is called _____ and is part of Indonesia.

7. Most of the low islands are made of _____ and barely rise above sea level.

8. Although the low islands support few trees besides the coconut palm, the high islands have dense tropical _____.

9. _____ covers about 98 percent of Antarctica's 5.4 million square miles.

10. Less precipitation falls in the _____ desert than in the Sahara in Africa.

Post-Reading Quick Check • After you have finished reading the section, in the space provided, describe the debate over mining in Antarctica.

_____

_____

_____

_____

_____

The Pacific Islands and Antarctica

EASTERN CHAPTER 26

SECTION 2

Reading the Section • As you read the section, answer each of the following questions in the space provided.

1. When did people begin settling the Pacific Islands? _____

2. Who was the first European to explore the Pacific? _____

3. What country took over Germany's Pacific territories after World War I? _____

4. What are three U.S. territories in the Pacific? _____

5. Which of the three Pacific Island regions is the most populous? _____

6. Where do nearly two thirds of all Pacific Islanders live? _____

7. What is Melanesia's largest city? _____

8. What religion do most Pacific Islanders practice? _____

9. How many islands make up Micronesia? _____

10. What is the largest Pacific region? _____

Post-Reading Quick Check • After you have finished reading the section, in the space provided, identify the three most important economic activities in the Pacific Islands today.

1. Economic activity: _____

2. Economic activity: _____

3. Economic activity: _____

The Pacific Islands and Antarctica

CHAPTER 33

EASTERN CHAPTER 26

SECTION 3

Reading the Section • As you read the section, examine each of the pairs of statements below. Circle the letter of the statement in each pair that is true.

1. **a.** In the 1970s British explorer James Cook sighted icebergs in the waters around Antarctica.
   b. In the 1770s British explorer James Cook sighted icebergs in the waters around Antarctica.

2. **a.** The first human expedition reached the North Pole in 1911.
   b. The first human expedition reached the South Pole in 1911.

3. **a.** The Antarctic Treaty of 1959 established Antarctica as a research site only.
   b. The Antarctic Treaty of 1959 established Antarctica as a military site only.

4. **a.** The United States has decided not to conduct research in Antarctica.
   b. The United States has research stations in Antarctica and at the South Pole.

5. **a.** Studies have found that carbon dioxide levels in the air have risen over time.
   b. Studies have found that carbon dioxide levels in the air have fallen over time.

6. **a.** Scientists have found that the ozone layer is intact around the world.
   b. Scientists have found a thinning in the ozone layer above Antarctica.

7. **a.** Antifreeze is a substance added to ice to keep the ice from turning to liquid.
   b. Antifreeze is a substance added to liquid to keep the liquid from turning to ice.

8. **a.** Antarctica is an excellent place to conduct research because many of the world's plants and animals live there.
   b. Antarctica is an excellent place to conduct research because humans have disturbed so little of the continent.

Post-Reading Quick Check • After you have finished reading the section, in the space provided, discuss the international agreement reached in 1991 concerning activities in Antarctica.

_____

_____

_____

_____

_____

People, Places, and Change

Answer Key

CHAPTER 1

Section 1 Activities

Reading the Section
Main Idea: Geographers view the world in a variety of ways. Topic I-B: People familiar with geography see relations between people, places, and environments. Topic II: Geographers study many issues. Topic II-B: Geographers study how governments change and how these changes affect people and their lives. Topic III-B: Regional geography divides the world into convenient parts for study. Topic III-C: Studying geography on the global level shows how events in one region are tied to events in other regions.

Post-Reading Quick Check
Answers will vary. Students may say that they hope the study of geography will help them to better understand the world around them and to better understand the relationships between people, places, and environments.

Section 2 Activities

Reading the Section
Location: focuses on absolute location and relative location; Place: describes places in terms of their physical or human characteristics; Human-Environment Interaction: focuses on how people and their surroundings affect each other; Movement: looks at how people migrate, or move, and how goods are traded; Region: focuses on areas of Earth's surface with one or more shared characteristics

Post-Reading Quick Check
1. People depend on their physical environment for survival.
2. People change their own behavior so that they are better suited to live in an environment.
3. People change the environment.

Section 3 Activities

Reading the Section
1. Cartography   2. Climatology
3. Meteorology   4. Human geography
5. Physical geography   6. Physical geography
7. Cartography   8. Human geography

Post-Reading Quick Check
Answers will vary. Students may mention grocery store managers, newspaper reporters, doctors, city managers, and emergency workers.

CHAPTER 2

Section 1 Activities

Reading the Section
1. b 2. b 3. a 4. a 5. b 6. b 7. a 8. a
9. a 10. b

Post-Reading Quick Check
1. atmosphere; the layer of gases that surrounds Earth
2. lithosphere; the solid, rocky outer layer of Earth
3. hydrosphere; all of Earth's water
4. biosphere; all plant and animal life

Section 2 Activities

Reading the Section
1. water vapor
2. the circulation of water from Earth's surface to the atmosphere and back
3. 97 percent
4. because they have few natural lakes
5. any smaller stream or river that flows into a larger stream or river
6. groundwater
7. Pacific, Atlantic, Indian, Arctic
8. in the gently sloping continental shelf
9. floods
10. to control flooding

Post-Reading Quick Check
Students should explain that the Sun drives the water cycle and should identify and describe evaporation, condensation, and precipitation.

Section 3 Activities

Reading the Section
1. d **2.** h **3.** f **4.** j **5.** a **6.** i **7.** c **8.** e
9. g **10.** b

Post-Reading Quick Check
1. colliding; can result in a very deep trench
2. moving apart; hot lava emerges from the gap that has been formed and builds a landform
3. sliding; can result in an earthquake

CHAPTER 3

Section 1 Activities

Reading the Section
1. Sun **2.** Tropic of Cancer **3.** stay the same **4.** greenhouse effect **5.** weight
6. barometer **7.** lighter; rises **8.** high; low
9. front **10.** ocean

Post-Reading Quick Check
1. prevailing winds; winds that blow in the same direction over large areas of Earth
2. trade winds; winds that blow in the subtropics
3. westerlies; winds that blow from the west in the middle latitudes
4. doldrums; area near the equator where few winds blow—warm air rises rather than blowing east or west

Section 2 Activities

Reading the Section
1. desert **2.** marine west coast **3.** ice cap
4. tropical savanna **5.** subarctic **6.** humid tropical **7.** humid subtropical **8.** tundra
9. humid continental **10.** steppe

Post-Reading Quick Check
weather: condition of the atmosphere in a local area for a short period of time; climate: weather in an area over a long period of time

Section 3 Activities

Reading the Section
1. extinct **2.** Ecology **3.** photosynthesis
4. Roots **5.** reproduce **6.** nutrients
7. communities **8.** ecosystem **9.** succession
10. humus

Post-Reading Quick Check
Plants are the basis for all food that animals eat. Some animals, say deer, eat only plants. When deer eat plants, they store some of the plant food energy in their bodies. Other animals, such as wolves, eat the deer and indirectly get the plant food eaten by the deer. The plants, deer, and wolves make up the food chain.

CHAPTER 4

Section 1 Activities

Reading the Section
Main Idea: Soil and forests are among the most important renewable resources on Earth. Topic I-B: Salt buildup is a threat to soil fertility. Topic I-D: Much farmland is lost to desertification and the expansion of cities and suburbs. Topic II: The forests provide both animals and people with food and shelter. Topic II-B: A great deal of forest is lost when it is cleared for farming, industry, and housing, and from industrial pollution.

Post-Reading Quick Check
Students may identify eight of the following: lumber, plywood, shingles, cellophane, furniture, plastics, fibers for cloth, fats, gums, medicines, nuts, oils, turpentine, waxes, rubber. Accept all reasonable answers.

Section 2 Activities

Reading the Section
1. semiarid **2.** aqueduct **3.** aquifer
4. desalinization **5.** desert plant **6.** acid rain **7.** ozone layer **8.** global warming

Post-Reading Quick Check

not having closed sewer systems, using too much chemical fertilizer and pesticide, dumping industrial waste

Section 3 Activities

Reading the Section

Metallic minerals: silver, iron, aluminum, steel, gold, mercury, copper, platinum; Nonmetallic minerals: sapphire, sulfur, ruby, quartz, talc, salt, emerald, diamond

Post-Reading Quick Check

1. substance that is part of Earth's crust
2. They are inorganic.
3. They occur naturally.
4. They are solids in crystalline form.
5. They have a definite chemical composition.

Section 4 Activities

Reading the Section

1. nonrenewable resources   2. coal   3. crude
4. Southwest Asia   5. Natural gas   6. hydro-electric power   7. 10   8. geothermal energy
9. solar energy   10. Chernobyl

Post-Reading Quick Check

Renewable resources can be replaced by Earth's natural processes. Nonrenewable resources, on the other hand, cannot be replaced by Earth's natural processes or are replaced very slowly.

CHAPTER 5

Section 1 Activities

Reading the Section

1. b   2. a   3. a   4. b   5. a   6. b   7. b
8. a

Post-Reading Quick Check

The term *symbol* refers to a sign that stands for something else. People learn symbols from their culture. For example, people who share a culture share the same sounds that make up a language, which might mean different things to people of other cultures. The

same is true of other types of symbols that people of a culture share.

Section 2 Activities

Reading the Section

1. demography   2. population density
3. Urbanization   4. birthrate; death rate
5. quaternary   6. gross national product
7. developed countries   8. democratic

Post-Reading Quick Check

Developing countries tend to be poorer than developed countries, and people often work in farming and other primary industries and earn low wages. Cities often are crowded with poorly educated people who have little access to health care or telecommunications.

Section 3 Activities

Reading the Section

Can—new fertilizers and special seeds allow more food to be grown; scientists probably will discover new energy sources; can make better use of existing resources; Cannot—amount of land available for farming is shrinking; shortage of fresh water; nonrenewable resources will run out; advancing pollution; food shortages and widespread disease

Post-Reading Quick Check

1. hinders economic growth
2. strains resources
3. lowers a country's ability to produce
4. must support a growing number of older people

CHAPTER 6

Section 1 Activities

Reading the Section

1. The Interior   2. The West   3. The East
4. The West   5. The Interior   6. The West
7. The West   8. The East   9. The Interior
10. The Interior

Post-Reading Quick Check

The United States has rich farmlands, producing a wide variety of crops. It also has valuable sources of coal, oil, natural gas, and minerals. Vast forests and offshore waters provide lumber, fish, and seafood. The natural beauty of the United States is another valuable resource, attracting tourists from around the world.

Section 2 Activities

Reading the Section
1. Asia **2.** Anasazi **3.** plantation **4.** Britain **5.** Hawaii **6.** U.S. Constitution **7.** bilingual **8.** July 4

Post-Reading Quick Check
1. Christmas **2.** Easter **3.** Hanukkah **4.** Kwanzaa

Section 3 Activities

Reading the Section
1. Interior West **2.** South **3.** Pacific **4.** Northeast **5.** Midwest **6.** Northeast **7.** Pacific **8.** Midwest **9.** South **10.** Interior West

Post-Reading Quick Check
1. peacekeeping **2.** trade **3.** poverty **4.** crime **5.** pollution **6.** urban sprawl

CHAPTER 7

Section 1 Activities

Reading the Section
1. United States **2.** Appalachian Mountains **3.** St. Lawrence River **4.** glaciers **5.** southwest **6.** tundra and ice cap **7.** about 50 percent **8.** coniferous forests **9.** paper **10.** minerals

Post-Reading Quick Check
Canada is a leading source of the world's nickel, zinc, and uranium. It also contains lead, copper, gold, silver, and coal. Saskatchewan has the world's largest deposits of potash. Alberta produces most of Canada's oil and natural gas.

Section 2 Activities

Reading the Section
1. c **2.** a **3.** a **4.** c **5.** d **6.** b **7.** b **8.** a

Post-Reading Quick Check
A mosaic is a picture made of tiny pieces of colored stone. Canada is a country of many different cultures that combine to form a single nation. These various cultures give Canada its identity.

Section 3 Activities

Reading the Section
1. h **2.** b **3.** f **4.** i **5.** d **6.** a **7.** j **8.** c **9.** e **10.** g

Post-Reading Quick Check
1. New Brunswick, Nova Scotia, Prince Edward Island
2. Quebec, Ontario
3. Manitoba, Saskatchewan, Alberta

CHAPTER 8

Section 1 Activities

Reading the Section
1. Texas **2.** Pacific; Mexico **3.** Oriental; Occidental **4.** Mexico City **5.** sinkholes **6.** tropics **7.** Summer **8.** Mexico City **9.** petroleum **10.** Silver

Post-Reading Quick Check
1. desert **2.** steppe **3.** savanna **4.** humid tropical

Section 2 Activities

Reading the Section
1. Maya **2.** Colonial Mexico **3.** Olmec **4.** Aztec **5.** Maya **6.** Aztec **7.** Colonial Mexico **8.** Olmec

Post-Reading Quick Check
In 1810 Miguel Hidalgo y Costilla began a revolt against Spanish rule, which continued until independence was achieved in 1821. In 1836 Texas broke away from Mexico, becoming part of the United States in 1845. A U.S.-Mexico argument over their common border

led to a war in which Mexico lost about half of its territory. In the early 1900s, many Mexicans grew unhappy with their government, and in 1910 the Mexican Revolution broke out.

Section 3 Activities

Reading the Section
Greater Mexico City—most developed and crowded region, includes national capital, heavily polluted, wealth and poverty exist side by side; Central Interior—lies north of the capital and extends toward both coasts, contains small towns, has fertile valleys and small family farms; Oil Coast—forested coastal plains between Tampico and Campeche along Gulf of Mexico, has oil production, farming, and ranching; Southern Mexico—Indian languages, poor with few cities and little industry, much unrest; Northern Mexico—dry region, prosperous and modern, cultural connections with the United States; The Yucatán—sparsely populated, farmers use slash-and-burn agriculture, ruins and sunny beaches

Post-Reading Quick Check
1. debts to foreign banks   **2.** high unemployment   **3.** inflation

CHAPTER 9

Section 1 Activities

Reading the Section
1. 20   **2.** Mountains   **3.** islands   **4.** Greater
5. Lesser   **6.** east   **7.** colliding   **8.** tropical savanna   **9.** winters; summers
10. Hurricanes

Post-Reading Quick Check
Agriculture is profitable in those areas of the region where volcanic ash has enriched the soil, and coffee, bananas, sugarcane, and cotton are important crops. The region is also important for timber and tourism. The region has few important energy and mineral resources, although Jamaica does have large reserves of bauxite and Panama has a huge copper deposit.

Section 2 Activities

Reading the Section
1. El Salvador   **2.** Nicaragua   **3.** Guatemala
4. Panama   **5.** Honduras   **6.** Costa Rica
7. El Salvador   **8.** Guatemala   **9.** Costa Rica   **10.** Belize   **11.** Nicaragua
12. Honduras

Post-Reading Quick Check
1. 3   **2.** 1   **3.** 4   **4.** 2

Section 3 Activities

Reading the Section
1. b   **2.** d   **3.** a   **4.** c   **5.** d   **6.** b   **7.** c   **8.** a
9. d   **10.** a

Post-Reading Quick Check
  1. calypso; Trinidad and Tobago
  2. reggae; Jamaica
  3. merengue; Dominican Republic

CHAPTER 10

Section 1 Activities

Reading the Section
Main Idea: Caribbean South America is an area of physical contrasts. Topic I-B: In the east, the Guiana Highlands have been eroding for centuries. Topic I-D: The Orinoco, the region's longest river, supports large ships and contains remarkable animals. Topic II-B: The climates of the Andes are divided by elevation into five zones. Topic II: Coffee is grown in the tierra templada, which has pleasant climates. Topic II-D: Grasslands and hardy shrubs grow in the paramo, which may have frost at any time. Topic II-E: The tierra helada, the zone of highest elevation, is always covered with snow. Topic III-A: The region is a rich agricultural area.

Post-Reading Quick Check
1. c   **2.** a   **3.** d   **4.** b

Section 2 Activities

Reading the Section
1. Chibcha   **2.** El Dorado   **3.** Spanish

4. Gran Colombia   **5.** Bogatá   **6.** coffee
7. cassava   **8.** *tejo*

Post-Reading Quick Check

1. oil   **2.** urban poverty, rapid population
growth   **3.** soccer, *tejo*   **4.** Roman
Catholicism

Section 3 Activities

Reading the Section

1. a   **2.** b   **3.** a   **4.** b   **5.** b   **6.** a   **7.** a   **8.** a
9. b   **10.** a

Post-Reading Quick Check

Students should discuss population, language, religion, music and dancing, and
sports.

Section 4 Activities

Reading the Section

1. Suriname   **2.** Suriname   **3.** French Guiana
4. Suriname   **5.** Guyana   **6.** French Guiana
7. French Guiana   **8.** Guyana   **9.** French
Guiana   **10.** Guyana

Post-Reading Quick Check

European countries had made slavery illegal
in the mid-1800s. Colonists in the Guianas
therefore needed a new source of labor to
work their plantations.

CHAPTER 11

Section 1 Activities

Reading the Section

Plains and Plateaus—Amazon River basin,
Brazilian Highlands, Brazilian Plateau, Gran
Chaco, Pampas, Patagonia, Tierra del Fuego;
Mountains—Andes; River Systems—Amazon,
Paraná; The Rain Forest—Amazon rain forest;
Climates—vary widely; Resources—resources
of the Amazon rain forest, gold, silver, cop-
per, iron, oil deposits, hydroelectric power

Post-Reading Quick Check

1. Brazil   **2.** Argentina   **3.** Uruguay
4. Paraguay

Section 2 Activities

Reading the Section

1. Asia   **2.** Portuguese   **3.** coffee   **4.** Roman
Catholic   **5.** samba   **6.** fifth   **7.** Amazon
8. northeast   **9.** southeast   **10.** Brasília

Post-Reading Quick Check

Brazil gained independence from Portugal in
1822. The country was ruled by an emperor
until 1889. Dictators and elected governments have ruled Brazil at various times
since then. Today, Brazil has an elected president and legislature.

Section 3 Activities

Reading the Section

1. land of silver   **2.** *encomienda*   **3.** gauchos
4. Falkland   **5.** Roman Catholic   **6.** *parillada*
7. Buenos Aires   **8.** Mercosur
The focus of Section 3 is Argentina.

Post-Reading Quick Check

1. Pampas   **2.** 1816   **3.** Spanish   **4.** tango
5. *milanga*   **6.** Pampas   **7.** 12   **8.** elected
president and legislature

Section 4 Activities

Reading the Section

1. U   **2.** U   **3.** P   **4.** B   **5.** P   **6.** U   **7.** P
8. B   **9.** P   **10.** U

Post-Reading Quick Check

1. An important energy source for Uruguay
   is hydroelectric power. One of the country's main challenges is developing the
   poor rural areas of the interior.
2. Paraguay is learning how to use its
   resources effectively. Consequently, its
   hydroelectric projects provide the country
   with more power than it needs, so the
   country sells the surplus power to Brazil
   and Argentina.

CHAPTER 12

Section 1 Activities

Reading the Section
1. b **2.** a **3.** a **4.** b **5.** b **6.** a **7.** a **8.** b
9. b **10.** b

Post-Reading Quick Check
El Niño conditions are caused by the buildup of warm water in the Pacific Ocean.

Section 2 Activities

Reading the Section
1. 4 **2.** 10 **3.** 7 **4.** 1 **5.** 2 **6.** 8 **7.** 3
8. 6 **9.** 5 **10.** 9

Post-Reading Quick Check
  1. walls without cement
  2. gold and silver objects
  3. huge irrigation projects
  4. stone-paved roads
  5. suspension bridges
  6. quipus

Section 3 Activities

Reading the Section
1. Bolivia **2.** Chile **3.** Ecuador **4.** Peru
5. Bolivia **6.** Chile **7.** Ecuador **8.** Peru

Post-Reading Quick Check
1. Quito **2.** La Paz and Sucre **3.** Lima
4. Santiago

CHAPTER 13 (EASTERN CHAPTER 6)

Section 1 Activities

Reading the Section
1. Greece **2.** Italy **3.** Portugal **4.** Spain
5. Portugal **6.** Italy **7.** Greece **8.** Greece
9. Spain **10.** Italy

Post-Reading Quick Check
Visitors are attracted to southern Europe by its sunny climate and natural beauty. There are also many castles, museums, ruins, and other cultural sites to see. In addition, Spain's beaches make it one of Europe's top tourist attractions.

Section 2 Activities

Reading the Section
1. a city and the land around it **2.** Athens
3. Alexander the Great **4.** Byzantine Empire
5. Greek Orthodox **6.** Ottoman Turks
7. agriculture **8.** about 40 percent
9. Athens **10.** Piraeus

Post-Reading Quick Check
Greece is famous for its beautiful sculpture, poetry, plays, pottery, and gold jewelry. The Greeks also made mosaics that were copied throughout Europe.

Section 3 Activities

Reading the Section
1. Latins **2.** aqueducts **3.** Byzantine Empire
4. Judaea **5.** pope **6.** Leonardo da Vinci
7. Amerigo Vespucci **8.** pizza
9. grapes **10.** Venice

Post-Reading Quick Check
  1. olives, bread, fish
  2. rice, butter, cheeses, mushrooms

Section 4 Activities

Reading the Section
1. Spain **2.** Latin **3.** Moors **4.** Isabella
5. England **6.** prime minister **7.** Spanish
8. Roman Catholic **9.** Lisbon **10.** Madrid

Post-Reading Quick Check
In the 1930s Spain's king lost power, and Spain became a workers' republic. The government tried to reduce the church's role and to give the nobles' land to ordinary farmers. A civil war resulted, with a victory for Franco's forces, and Franco ruled Spain until 1975. Today Spain is a democracy.

CHAPTER 14 (Eastern Chapter 7)

Section 1 Activities

Reading the Section
Main Idea: Nearness to the sea has affected west-central Europe in a number of ways. Topic I-B: West-central Europe contains areas of uplands. Topic II: West-central Europe has several climates and a number of waterways. Topic II-B: West-central Europe has many navigable rivers important for trade and travel. Topic III-B: Energy resources are generally in short supply. Topic III-C: Alternative sources of energy help supply the region's energy needs.

Post-Reading Quick Check
1. Seine **2.** Loire **3.** Garonne **4.** Rhone **5.** Rhine **6.** Danube **7.** Elbe **8.** Oder **9.** Weser

Section 2 Activities

Reading the Section
1. c **2.** d **3.** b **4.** b **5.** d **6.** d **7.** c **8.** a

Post-Reading Quick Check
Bastille Day, celebrated on July 14, marks the day in 1789 when a Paris mob stormed a royal prison, the Bastille. The French consider this day the start of the French Revolution. They celebrate the day with fireworks displays and dancing in the streets.

Section 3 Activities

Reading the Section
1. a **2.** a **3.** b **4.** b **5.** b **6.** a **7.** b **8.** a **9.** b

Post-Reading Quick Check
East and West Germany were reunited in 1990. In 1999 the capital was moved from Bonn to Berlin. The German government is now a republic with a parliament who elect the president and prime minister, or chancellor. Germany also is a member of the EU and NATO.

Section 4 Activities

Reading the Section
1. Holland **2.** Spanish **3.** France **4.** Belgium **5.** parliament **6.** Dutch **7.** dairy **8.** cookies **9.** tulips **10.** banking

Post-Reading Quick Check
The people of Luxembourg and Belgium are mostly Roman Catholic. The Netherlands is more evenly divided among Catholics, Protestants, and those who have no religious ties.

Section 5 Activities

Reading the Section
1. S **2.** A **3.** S **4.** S **5.** S **6.** A **7.** A **8.** A **9.** S **10.** A

Post-Reading Quick Check
1. 1863
2. Switzerland
3. to help relieve wartime human suffering

CHAPTER 15 (Eastern Chapter 8)

Section 1 Activities

Reading the Section
1. i **2.** a **3.** g **4.** d **5.** j **6.** c **7.** e **8.** h **9.** f **10.** b

Post-Reading Quick Check
Most of Europe's original forests were cleared centuries ago. Sweden and Finland, though, still have large, timber-producing coniferous forests. Farmers in the region grow many kinds of cool-climate crops. The farms in northern Europe are among the world's most productive farms.

Section 2 Activities

Reading the Section
1. Normans **2.** one fourth **3.** coal **4.** parliament **5.** 60 **6.** English **7.** potatoes **8.** queen's **9.** London **10.** Religion

Post-Reading Quick Check

The farms in Britain produce about 60 percent of the country's food, even though only about 1 percent of the labor force works in agriculture. The country's important products include grains, potatoes, vegetables, and meat.

Section 3 Activities

Reading the Section

Government—president, parliament, prime minister; Culture—English-speaking, Gaelic League, Gaelic and English used in schools and in official documents, folk dancing and music, mainly Roman Catholic; Economy—industrial country, low taxes, foreign companies, EU membership; Cities—Dublin, Cork, Galway

Post-Reading Quick Check

1. conquest by the British and religious differences with the British
2. potato famine led to starvation

Section 4 Activities

Reading the Section

1. Denmark   2. Greenland   3. Norway
4. Finland   5. Sweden   6. Lapland
7. Norway   8. Iceland   9. Norway
10. Denmark   11. Finland   12. Sweden

Post-Reading Quick Check

1. The languages are closely related.
2. Most of the people are Protestant.
3. All have democratic governments.

CHAPTER 16 (EASTERN CHAPTER 9)

Section 1 Activities

Reading the Section

1. Estonia, Latvia, Lithuania
2. Danube
3. Balkan Peninsula
4. Romania
5. Danube River
6. about 600

7. long, snowy winters and short, rainy summers
8. layered rock that yields oil when heated
9. Slovakia, Slovenia
10. amber

Post-Reading Quick Check

earlier communist rule; industrial development was considered more important than the environment

Section 2 Activities

Reading the Section

1. Velvet Revolution   2. Genghis Khan
3. Finnish   4. Latvia   5. Lithuania
6. Poland   7. Czechoslovakia   8. Czech Republic   9. Slovakia   10. Hungary

Post-Reading Quick Check

1. Tallinn   2. Riga   3. Vilnius   4. Prague
5. Bratislava   6. Budapest

Section 3 Activities

Reading the Section

1. Ottoman Turks   2. Serbia   3. Austria
4. Belgrade   5. Bosnia   6. Macedonia
7. agriculture   8. market economy

Post-Reading Quick Check

combine foods of the Hungarians and Slavs with those of Greeks, Turks, and Italians; yogurt, soft cheeses, fruits, nuts, vegetables, roast goat or lamb

CHAPTER 17 (EASTERN CHAPTER 10)

Section 1 Activities

Reading the Section

1. climate and vegetation
2. physical features
3. resources
4. resources
5. physical features
6. climate and vegetation
7. resources
8. physical features
9. physical features

10. climate and vegetation

Post-Reading Quick Check

1. tundra **2.** subarctic **3.** humid continental
4. steppe

Section 2 Activities

Reading the Section

1. Soviet Union
2. Russian Empire
3. Russia Today
4. Early Russia
5. Soviet Union
6. Early Russia
7. Russia Today
8. Russian Empire

Post-Reading Quick Check
The Mongols demanded taxes but ruled the region through local leaders. Over time, the local leaders established various states.

Section 3 Activities

Reading the Section
Moscow—Russia's capital and largest city, linked to all points in Russia, Kremlin, industrial area; St. Petersburg—second-largest city, former capital and home to czars, few natural resources, major trade, industrial, and educational center; Volga—major shipping route, hydroelectric power, coal and oil deposits, heavy-machine industries and giant factories; Urals—mining, smelters, manufacturing

Post-Reading Quick Check

1. Russian nation expanded outward from here
2. contains the bulk of the Russian population
3. contains the national capital and large industries
4. country's most productive farming region

Section 4 Activities

Reading the Section

1. more than 5 million square miles
2. Arctic Ocean
3. Mongolia, China

4. "Sleeping Land"
5. long, dark, severe, little snow
6. ethnic Russians
7. Trans-Siberian Railroad, Baikal-Amur Mainline
8. timber, mineral ores, diamonds, coal, oil, natural gas
9. lumbering, mining
10. Kuzbas
11. Novosibirsk
12. Lake Baikal

Post-Reading Quick Check
Many people have been worried that pollution from a nearby paper factory and other development threaten the species that live in and around the lake. Recently, scientists and others have proposed plans that allow for economic development but that protect the environment.

Section 5 Activities

Reading the Section

1. a **2.** a **3.** b **4.** a **5.** b **6.** b **7.** a **8.** a

Post-Reading Quick Check
Russia and Japan have argued about ownership of Sakhalin and the Kurils since the 1850s. At times they have been divided between Japan and Russia or the Soviet Union. The Soviets took control of all the islands after WW II. However, Japan still claims rights to the southernmost islands.

CHAPTER 18 (EASTERN CHAPTER 11)

Section 1 Activities

Reading the Section

1. Russia **2.** Armenia **3.** Carpathian
4. Black **5.** Elbrus **6.** Dnieper **7.** Farmland
8. reserves **9.** humid continental **10.** gas

Post-Reading Quick Check
The Ukraine's rich farmlands have been the country's greatest natural resource for centuries. Farming is also important in Belarus. In addition, lowland areas of the Caucasus have rich soil and good conditions for growing crops.

People, Places, and Change

Section 2 Activities

Reading the Section
1. U **2.** UB **3.** B **4.** U **5.** B **6.** U **7.** B
8. UB **9.** U **10.** B

Post-Reading Quick Check
1. 5 **2.** 7 **3.** 4 **4.** 8 **5.** 3 **6.** 1 **7.** 6 **8.** 2

Section 3 Activities

Reading the Section
1. Armenia **2.** Georgia **3.** Azerbaijan
4. Armenia **5.** Azerbaijan **6.** Armenia
7. Georgia **8.** Azerbaijan **9.** Azerbaijan
10. Georgia

Post-Reading Quick Check
1. civil war
2. ethnic minorities who want independence
3. disagreements over gas and oil rights

CHAPTER 19 (EASTERN CHAPTER 12)

Section 1 Activities

Reading the Section
1. d **2.** a **3.** b **4.** e **5.** a **6.** b **7.** c **8.** a
9. e **10.** c

Post-Reading Quick Check
Uzbekistan, Kazakhstan, and Turkmenistan all have huge oil and natural gas deposits. Several countries in Central Asia are also rich in other minerals, with deposits of gold, copper, uranium, zinc, and lead. Kazakhstan has large deposits of coal. Rivers in Kyrgystan and Tajikistan could be used to create hydroelectric power.

Section 2 Activities

Reading the Section
1. through Central Asia
2. for protection
3. Europeans discovered they could sail to East Asia through the Indian Ocean.
4. Genghis Khan and the Mongols
5. following the Russian Revolution
6. Russian

7. when the Soviet Union broke up in 1991
8. Latin
9. centralized, Soviet-style
10. food processing, cloth-making, mining, and pumping oil

Post-Reading Quick Check
1. schools
2. hospitals
3. more freedom for women

Section 3 Activities

Reading the Section
1. Uzbekistan **2.** Kyrgyzstan
3. Kyrgyzstan **4.** Kyrgyzstan **5.** Tajikistan
6. Kazakhstan **7.** Tajikistan **8.** Uzbekistan
9. Kazakhstan **10.** Turkmenistan

Post-Reading Quick Check
The Soviet-style government fought against a mixed group of reformers. Some reformers wanted democracy; others wanted government by Islamic law. Although a peace agreement was signed in 1997, tensions remain high.

CHAPTER 20 (EASTERN CHAPTER 13)

Section 1 Activities

Reading the Section
1. g **2.** f **3.** i **4.** e **5.** a **6.** h **7.** j **8.** c
9. d **10.** b

Post-Reading Quick Check
The most important mineral resource of the region is oil. With the exception of Iran, the region's countries are not rich in other resources. Iran, however, has deposits of many different metals.

Section 2 Activities

Reading the Section
Islam—Muhammad, Muslims, Sunni, Shia, Qur'an; Government—Saud family, monarchy, Riyadh; Economy—oil, OPEC, imported food; People—ethnic Arabs, Arabic language, city-dwellers, sizable middle class, free health care and education, limited freedom for

women; Customs—pray five times a day, Friday holy day, modesty, two major celebrations

Post-Reading Quick Check

Students should identify Kuwait, Bahrain, Qatar, the United Arab Emirates, Oman, and Yemen.

Section 3 Activities

Reading the Section

1. 7　**2.** 8　**3.** 6　**4.** 2　**5.** 1　**6.** 4　**7.** 9　**8.** 3　**9.** 10　**10.** 5

Post-Reading Quick Check

The Iran-Iraq War and the Persian Gulf War damaged Iraq's oil industry. In addition, the UN placed an embargo on Iraq.

Section 4 Activities

Reading the Section

1. A　**2.** A　**3.** I　**4.** I　**5.** A　**6.** A　**7.** I　**8.** A　**9.** I　**10.** I

Post-Reading Quick Check

1. Oil is the main industry in Iran, which is a member of OPEC. Its other industries include construction, food processing, and carpet production. Farming also is important, with about a third of the workforce employed in agriculture.
2. War has severely damaged Afghanistan's industry, trade, and transportation systems. Farming and herding are the most important economic activities there now.

CHAPTER 21 (EASTERN CHAPTER 14)

Section 1 Activities

Reading the Section

1. b　**2.** b　**3.** a　**4.** b　**5.** a　**6.** a　**7.** a　**8.** a

Post-Reading Quick Check

Students may suggest that the Dead Sea is so salty that nothing can live in it.

Section 2 Activities

Reading the Section

1. Alexander the Great
2. Constantinople
3. losing
4. Latin
5. National Assembly
6. Muslim
7. only one wife
8. Making clothing
9. Kurds
10. food

Post-Reading Quick Check

Many of the people in Turkey follow Islamic principles, which tend to be more traditional. These people do not approve of the modernization efforts because they are contrary to Islamic principles.

Section 3 Activities

Reading the Section

1. 6　**2.** 4　**3.** 8　**4.** 7　**5.** 10　**6.** 1　**7.** 2　**8.** 5　**9.** 9　**10.** 3

Post-Reading Quick Check

1. Gaza Strip; small, crowded piece of coastal land; more than a million Palestinians live there; poor with almost no resources
2. Golan Heights; hilly area on the Syrian border
3. West Bank; largest of the occupied areas; thousands of Jews moved there

Section 4 Activities

Reading the Section

1. Jordan　**2.** Syria　**3.** Jordan　**4.** Lebanon　**5.** Syria　**6.** Lebanon　**7.** Jordan　**8.** Lebanon　**9.** Lebanon　**10.** Syria

Post-Reading Quick Check

1. Damascus　**2.** Beirut　**3.** Amman

CHAPTER 22 (EASTERN CHAPTER 15)

Section 1 Activities

Reading the Section
Main Idea: The physical geography of North Africa includes physical features, climate, and resources. Topic I-B: The Sahara also includes mountains and depressions. Topic I-D: East of the Nile is the Sinai Peninsula and the Suez Canal. Topic II: North Africa has three main climates and many plants and animals. Topic II-B: Much of the northern coast west of Egypt has a Mediterranean climate, with grasses, shrubs, and forests. Topic II-C: Between the Mediterranean climate and the Sahara is a steppe climate, with scrubs and grasses. Topic III-B: Oil and gas are important resources.

Post-Reading Quick Check
The annual flooding left tons of nutrient-rich silt in surrounding fields. Silt makes good soil for growing crops.

Section 2 Activities

Reading the Section
1. early Egyptians **2.** hieroglyphs **3.** British **4.** Israel **5.** Bedouins **6.** Egypt **7.** couscous **8.** Muhammad

Post-Reading Quick Check
Beginning in the A.D. 600s, Arab armies from Southwest Asia swept across North Africa. They brought Islam and the Arabic language with them to the region.

Section 3 Activities

Reading the Section
Rural Egypt—more than half of all Egyptians live in small villages and other rural areas; most are *fellahin;* Cities—Cairo, the capital, and Alexandria; Industries—textiles, tourism, oil; Suez Canal—source of income from tolls; Agriculture—cotton, vegetables, grain, fruit; Challenges—need for fertilizers; divided over country's role in the world; lack of clean water and spread of disease; role of Islam in the country

Post-Reading Quick Check
1. work large farms owned by powerful families
2. receive money from family members working abroad
3. work in Europe or oil-rich Southwest Asian countries

Section 4 Activities

Reading the Section
1. Tunisia **2.** Morocco **3.** Libya **4.** Morocco **5.** Tunisia **6.** Algeria **7.** Libya **8.** Algeria **9.** Libya **10.** Algeria

Post-Reading Quick Check
1. d **2.** b **3.** c **4.** a

CHAPTER 23 (EASTERN CHAPTER 16)

Section 1 Activities

Reading the Section
1. Sahara
2. northernmost parts of region
3. desert
4. Sahel
5. south of the Sahara
6. steppe
7. savanna
8. farther south of the Sahara
9. savanna
10. coast and forest
11. along the Atlantic Ocean and Gulf of Guinea coasts
12. humid tropical

Post-Reading Quick Check
The Niger bring life-giving water to the area. In the Sahel, it divides into many channels, swamps, and lakes, where a large variety of animals live.

Section 2 Activities

Reading the Section
1. Archaeology **2.** Ghana **3.** Mansa Mūsā **4.** Timbuktu **5.** Gold Coast **6.** Africans

7. Liberia   **8.** Portugal   **9.** Islam   **10.** circular

Post-Reading Quick Check

1. civil wars and military rulers
2. high birthrates that result in too few jobs for too many people
3. inability of many people to afford education for their children

Section 3 Activities

Reading the Section

1. Mali   **2.** Burkina Faso   **3.** Chad
4. Mauritania   **5.** Niger   **6.** Mali   **7.** Burkina Faso   **8.** Niger   **9.** Chad   **10.** Mauritania

Post-Reading Quick Check

Students may discuss former colonial status, language, religion, poverty, the expanding desert, and drought.

Section 4 Activities

Reading the Section

1. d   **2.** h   **3.** f   **4.** a   **5.** i   **6.** c   **7.** g   **8.** b   **9.** j   **10.** e

Post-Reading Quick Check

1. both have been troubled by unstable government since independence
2. both are poor
3. people of both depend on farming and herding
4. main crops of both are palm tree products, cacao, and coffee
5. both have experienced periods of military rule

CHAPTER 24 (EASTERN CHAPTER 17)

Section 1 Activities

Reading the Section

1. a   **2.** a   **3.** a   **4.** b   **5.** a   **6.** a   **7.** b   **8.** b   **9.** a   **10.** a

Post-Reading Quick Check

1. long, deep valley with mountains or plateaus on either side
2. formed when Earth's tectonic plates move away from each other

Section 2 Activities

Reading the Section

1. Christianity   **2.** Portuguese   **3.** British
4. Kenya   **5.** Ethiopia   **6.** 1960s   **7.** longest
8. language, spoken   **9.** Christians
10. Burundi

Post-Reading Quick Check

Islam was brought to East Africa by Arabic-speaking nomads who spread slowly into northern Sudan from Egypt. Islam also spread to the coastal region of what is now Somalia.

Section 3 Activities

Reading the Section

1. British   **2.** 1960s   **3.** Tanganyika, Zanzibar
4. Olduvai Gorge   **5.** Tanzania   **6.** Belgium
7. Tutsi, Hutu   **8.** Uganda   **9.** Sudan
10. Sahara

Post-Reading Quick Check

1. viewed land as a sign of personal wealth, power, and prosperity
2. viewed land as important for the amount of food it could grow

Section 4 Activities

Reading the Section

1. b   **2.** c   **3.** b   **4.** a   **5.** c   **6.** d   **7.** a   **8.** c   **9.** d   **10.** d

Post-Reading Quick Check

It looks like a rhinoceros horn pointing at the Arabian Peninsula.

CHAPTER 25 (EASTERN CHAPTER 18)

Section 1 Activities

Reading the Section

1. F   **2.** T   **3.** T   **4.** F   **5.** T   **6.** T   **7.** F   **8.** F   **9.** T   **10.** T   **11.** T   **12.** F

Post-Reading Quick Check

1. Congo River; flows westward to the Atlantic Ocean; hundreds of smaller rivers flow into it

People, Places, and Change

2. Zambezi River; flows eastward to the Indian Ocean; famous for its falls, hydro-electric dams, and lakes

Section 2 Activities

Reading the Section
1. a **2.** b **3.** c **4.** a **5.** b **6.** c **7.** d **8.** d **9.** c **10.** c

Post-Reading Quick Check
1. putting an end to the wars
2. developing natural resources
3. stopping the spread of disease

Section 3 Activities

Reading the Section
1. 7 **2.** 4 **3.** 1 **4.** 6 **5.** 2 **6.** 8 **7.** 3 **8.** 5

Post-Reading Quick Check
1. more than 50 million
2. more than 200
3. French
4. Kinshasa
5. copper, gold, diamonds, cobalt, tropical rain forest
6. peace, stable government, better schools and health care, road and railroad repair and expansion

Section 4 Activities

Reading the Section
1. b **2.** a **3.** a **4.** b **5.** a **6.** a **7.** b **8.** b **9.** a **10.** a

Post-Reading Quick Check
After it won independence, Angola plunged into civil war. Fighting has continued on and off since then.

CHAPTER 26 (EASTERN CHAPTER 19)

Section 1 Activities

Reading the Section
Countries—Namibia, South Africa, Mozambique, Botswana, Zimbabwe, Lesotho, Swaziland; Landforms—large interior plateau, Drakensberg, veld; Climate—from desert to cool uplands; Deserts & Rivers—Kalahari, Namib, Orange River, Augrabies Falls, Limpopo; Resources—gold, diamonds, platinum, copper, uranium, coal, and iron ore

Post-Reading Quick Check
1. Botswana **2.** Zimbabwe **3.** Lesotho **4.** Swaziland

Section 2 Activities

Reading the Section
1. 4 **2.** 3 **3.** 7 **4.** 1 **5.** 6 **6.** 5 **7.** 8 **8.** 10 **9.** 9 **10.** 2

Post-Reading Quick Check
The main language of the colony set up by the Dutch at Cape Town was Dutch. Over time, Khoisan, Bantu, and Malay words were added. This created a new language called Afrikaans.

Section 3 Activities

Reading the Section
Main Idea: Despite racist laws, South Africa has been southern Africa's most important country in terms of population and resources. Topic I-B: South Africa set up apartheid after WW II. Topic I-D: Nonwhites had few rights, and those who protested were sent to prison. Topic II: People around the world protested apartheid. Topic II-B: In the late 1980s South Africa began to move away from apartheid. Topic III-B: Most industries and good farmland are still owned by whites. Topic III-C: New problems have arisen since the end of apartheid.

Post-Reading Quick Check
1. banned trade with South Africa
2. refused to invest money in South Africa
3. refused to include South Africa in meetings and competitions

Section 4 Activities

Reading the Section
1. Zimbabwe **2.** Namibia **3.** Madagascar **4.** Botswana **5.** Botswana **6.** Mozambique

7. Zimbabwe 8. Namibia 9. Mozambique
10. Madagascar

Post-Reading Quick Check
1. b 2. d 3. a 4. c

CHAPTER 27 (EASTERN CHAPTER 20)

Section 1 Activities

Reading the Section
1. China 2. Taiwan 3. Himalayas
4. Plateau 5. Gobi 6. North 7. Yellow
8. Chang 9. coal 10. Taiwan

Post-Reading Quick Check
The southeastern coastal region of China is
the most humid. Moving northwestward, the
climate becomes drier and drier. In the
extreme northwest the climate is a true
desert.

Section 2 Activities

Reading the Section
1. Huang He 2. Qin 3. compass
4. Mongols 5. Sun Yat-sen 6. Taiwan
7. Cultural 8. Mandarin 9. Confucius
10. Buddhism

Post-Reading Quick Check
turned all private land into government-run
farms; factories and their production put
under government control; government
owned all housing and told people where to
live; separated or relocated families; women
given equal status and duties; restricted family size; prohibited all religious worship

Section 3 Activities

Reading the Section
1. China
2. 10 percent
3. 40
4. Beijing
5. Shanghai
6. Hong Kong
7. command economy
8. 10 percent

9. wheat, sorghum
10. trade advantages from the United States

Post-Reading Quick Check
1. has a huge agricultural workforce
2. farmers create new farmland by cutting
   terraces into hillsides

Section 4 Activities

Reading the Section
1. M 2. T 3. T 4. M 5. T 6. M 7. T
8. T 9. M 10. M

Post-Reading Quick Check
Taiwan's government began encouraging
industrial development during the 1950s.
Taiwan now has one of Asia's most success-
ful economies. It is a world leader in the production and export of computers and sports
equipment.

CHAPTER 28 (EASTERN CHAPTER 21)

Section 1 Activities

Reading the Section
Location—Korean Peninsula about 600 miles
(965 km) from mainland Asia; Mountains—
common in all three countries, most of
Japan's mountains formed by volcanic activity; Ring of Fire—Japan lies along subduction
zone and has active volcanoes and earthquakes, Korea not located on subduction
zone and has no active volcanoes so earthquakes are rare; Natural Resources—only
North Korea rich in resources, Koreas suitable
for producing hydroelectric power, Japan has
rich supply of fish; Climate Types—humid
continental and humid subtropical

Post-Reading Quick Check
1. Oyashio Current; dives beneath the
   warm, less dense Japan Current and
   keeps areas near the coast of northern
   Japan cool in summer
2. Japan Current; flows northward from
   the tropical North Pacific and warms
   southern Japan

Section 2 Activities

Reading the Section
1. 9 **2.** 5 **3.** 8 **4.** 2 **5.** 10 **6.** 12 **7.** 11
8. 3 **9.** 7 **10.** 1 **11.** 6 **12.** 4

Post-Reading Quick Check
1. Shintoism; centers around the spirits of natural places, sacred animals, and ancestors
2. Buddhism; introduced from China, with shrines often located next to older *kami* shrines
3. Confucianism; introduced from China, principles include respect for elders, parents, and rulers
4. Christianity; introduced in Japan by Spanish and Dutch

Section 3 Activities

Reading the Section
1. most densely
2. Only about 11 percent
3. 30 million
4. Tokyo
5. Over 99 percent
6. futons
7. nuclear power
8. fishing industry
9. smaller
10. terracing

Post-Reading Quick Check
1. Osaka, has been a trading center for centuries
2. Kobe; important seaport
3. Kyoto; Japan's capital for more than 1,000 years

Section 4 Activities

Reading the Section
1. early modern Korea
2. ancient Korea
3. early modern Korea
4. Korea since World War II
5. early modern Korea
6. ancient China
7. Korea since World War II
8. ancient Korea
9. Korea since World War II
10. early modern Korea

Post-Reading Quick Check
At the end of World War II, U.S. and Soviet troops arrived in Korea to oversee the departure of the Japanese. The Soviets helped communist leaders take power in the north, while the United States supported a democracy in the south. Unable to agree on a plan to reunite, the south in 1948 officially became the Republic of Korea, and the north became the Democratic People's Republic of Korea.

Section 5 Activities

Reading the Section
1. Seoul
2. multiparty democratic government
3. Christianity
4. *chaebol*
5. rice
6. kimchi
7. Communist Party
8. cooperatives

Post-Reading Quick Check
North Korea has only outdated technology, so unlike South Korea it cannot produce the high-quality goods needed to compete in the international market.

CHAPTER 29 (EASTERN CHAPTER 22)

Section 1 Activities

Reading the Section
Main Idea: Southeast Asia includes a large mainland region, thousands of large and small islands, and great rivers. Topic I-B: The archipelagos are the Malay Archipelago and the Philippines. Topic I-D: Five river systems drain the mainland, and the greatest river is the Mekong. Topic II: Climate, vegetation, and wildlife. Topic II-B: The region is home to thousands of different plants, many of which are endangered. Topic III-B: The

region's rich volcanic soils and floodplains are good for farming. Topic III-C: Crops include rice, coconuts, palm oil, sugarcane, coffee, and spices.

Post-Reading Quick Check
The plants and animals are endangered because the rain forests are being cleared for farmland, tropical wood, and mining. Some governments in the region are taking steps to save their rain forests.

Section 2 Activities

Reading the Section
1. 5 **2.** 9 **3.** 6 **4.** 2 **5.** 8 **6.** 3 **7.** 10 **8.** 1
9. 7 **10.** 4

Post-Reading Quick Check
1. Hinduism **2.** Buddhism **3.** Islam
4. Indonesia **5.** Roman Catholicism **6.** rice
7. coconut

Section 3 Activities

Reading the Section
1. Laos **2.** Myanmar **3.** Thailand
4. Vietnam **5.** Myanmar **6.** Cambodia
7. Vietnam **8.** Thailand **9.** Cambodia
10. Laos

Post-Reading Quick Check
Many Southeast Asians are moving to the cities to look for work, The cities have many, businesses, services, and opportunities that are not found in rural areas.

Section 4 Activities

Reading the Section
1. d **2.** b **3.** e **4.** a **5.** c **6.** b **7.** e **8.** c
9. d **10.** a

Post-Reading Quick Check
Crime rates in Singapore are low because the government is very strict. For example, there are stiff fines for littering, and people caught transporting illegal drugs can be executed. The government even bans chewing gum and certain movies and music.

CHAPTER 30 (EASTERN CHAPTER 23)

Section 1 Activities

Reading the Section
  1. a very large landmass that is smaller than a continent
  2. Himalayas
  3. created when two tectonic plates collided
  4. Eastern Ghats, Western Ghats
  5. Ganges
  6. the "Mother River"
  7. Thar Desert
  8. humid tropical
  9. agriculture
  10. iron ore, bauxite, uranium, coal

Post-Reading Quick Check
1. Himalayas **2.** Gangetic Plain **3.** Deccan

Section 2 Activities

Reading the Section
1. Indus **2.** Indo-Aryans **3.** sultanate
4. Babur **5.** Akbar **6.** Shah Jahan **7.** British
8. sepoys **9.** Gandhi **10.** India, Pakistan

Post-Reading Quick Check
Gandhi used nonviolent mass protest. He called for Indians to peacefully refuse to cooperate with the British. Gandhi led protest marches and urged Indians to boycott British goods. He also went on hunger strikes.

Section 3 Activities

Reading the Section
Caste System—groups of people born into their positions in society, from highest status to lowest status; after independence, government did away with system; Languages—24 languages; Hindi official national language; Government—democratic; world's largest democracy; Economy—mixture of traditional and modern; 65 percent are farmers; green revolution; among the world's top 10 industrial countries

Post-Reading Quick Check
1. Hinduism **2.** Buddhism **3.** Jainism
4. Sikhism

People, Places, and Change

CHAPTER 31 (EASTERN CHAPTER 24)

Section 1 Activities

Reading the Section
1. b **2.** a **3.** b **4.** a **5.** a **6.** b **7.** a **8.** a
9. b **10.** b

Post-Reading Quick Check
In the early and late weeks of the monsoon season, cyclones sweep in from the Bay of Bengal. These cyclones bring high winds and heavy rains, often accompanied by storm surges. The summer monsoon brings hot, wet weather to the lowland areas of Bhutan and Nepal.

Section 2 Activities

Reading the Section
1. Indus **2.** religion **3.** Bangladesh
4. Islamic **5.** highest **6.** Bengal **7.** Muslims **8.** Dhaka **9.** parents **10.** Bengali

Post-Reading Quick Check
Runoff from heavy rains in the Himalayas causes extensive flooding. When the land is flooded, sewage often washes into the water. As a result, the people in Bangladesh suffer from epidemics of diseases, such as cholera.

Section 3 Activities

Reading the Section
1. b **2.** d **3.** a **4.** c **5.** b **6.** a **7.** d **8.** a
9. c **10.** b

Post-Reading Quick Check
1. Kathmandu **2.** Thimphu **3.** Colombo
4. Male

CHAPTER 32 (EASTERN CHAPTER 25)

Section 1 Activities

Reading the Section
1. Australia **2.** Great Dividing Range
3. Great Barrier Reef **4.** eucalyptus
5. Aborigines **6.** British **7.** Canberra
8. rugby **9.** Sydney **10.** bush

Post-Reading Quick Check
1. animal that carries its young in a pouch
2. kangaroo
3. koala

Section 2 Activities

Reading the Section
1. Southern Alps **2.** kiwi **3.** Maori
4. Great Britain **5.** Wellington **6.** *whare whakairo* **7.** Auckland **8.** Christchurch
9. wool **10.** United States

Post-Reading Quick Check
Most of the people who live in New Zealand have European backgrounds, although Maori make up 10 percent of the population. In addition, there are a growing number of ethnic Asians and people from other Pacific islands. Most New Zealanders speak English and are Christian. Sports are popular there, and Maori culture continues to be popular.

CHAPTER 33 (EASTERN CHAPTER 26)

Section 1 Activities

Reading the Section
1. one third **2.** Micronesia **3.** oceanic
4. rock **5.** New Guinea **6.** Irian Jaya
7. coral **8.** rain forests **9.** ice **10.** polar

Post-Reading Quick Check
Antarctica has many mineral resources, but there is debate over the mining of those resources. Some people worry that mining would harm the environment. Others questions whether mining would even be worthwhile for businesses.

Section 2 Activities

Reading the Section
1. at least 40,000 years ago
2. Ferdinand Magellan
3. Japan
4. Northern Mariana Islands, Guam, Wake Island
5. Melanesia

6. Papua New Guinea
7. Port Moresby
8. Christianity
9. more than 2,000
10. Polynesia

Post-Reading Quick Check
1. tourism **2.** agriculture **3.** fishing

Section 3 Activities

Reading the Section
1. b **2.** b **3.** a **4.** b **5.** a **6.** b **7.** b **8.** b

Post-Reading Quick Check
The 1991 agreement bans most activities in Antarctica that do not have a scientific purpose. It bans mining and drilling and limits tourism.